The Use of Algae in Human Health

First published 2025 in Great Britain and the United States by ISTE Ltd and John Wiley & Sons, Inc.

ISTE Ltd
27-37 St George's Road
London SW19 4EU
UK

www.iste.co.uk

John Wiley & Sons, Inc.
111 River Street
Hoboken, NJ 07030
USA

www.wiley.com

Library of Congress Control Number: 2024951144

British Library Cataloguing-in-Publication Data
A CIP record for this book is available from the British Library
ISBN 978-1-78630-956-3

Contents

on the use of plants or their extracts, mainly administered in the form of decoctions. Seaweed is one of the plants used in traditional medicine, particularly in Asia. Numerous examples of the use of these marine plants are described in Chinese, Japanese and Vietnamese traditional medicine. In traditional Chinese medicine, seaweed-based preparations can be prescribed for the treatment of goiters and various respiratory ailments such as bronchitis. But it is not just Asian folk medicine that recommends the use of seaweed. Indeed, seaweed is also used in traditional medicine in Africa (South Africa), South America (Chile) and the Pacific islands.

Once used empirically, traditional medicines are now able to rely on advances in the knowledge of active principles or algal molecules with interesting properties for human health. Indeed, the therapeutic properties concerned relate to the presence of antioxidant, anti-inflammatory, antibacterial, antiviral, antithrombotic and even antitumoral activities.

This book offers a historical and current overview of traditional medicines that integrate seaweed (macroalgae) into curative or preventive practices in human health. It also lists the activities of therapeutic interest associated with algae and described in the scientific literature. Where possible, it also reviews the algal compounds identified as biochemical supports for pharmacological activities.

Finally, the book also describes the limits to the development of algal molecules for new drugs.

December 2024

Introduction

The Sumerian poem *The Epic of Gilgamesh* is considered one of the oldest texts in human history (2,100 BCE). In this mythical text, the demigod Gilgamesh sets out on a quest to discover the secret of immortality. On his journey, the wise Utnapisthim advises Gilgamesh to search the bottom of the sea for a thorny plant or herb of youth that has the power to rejuvenate any person or animal that consumes it (O'Connor 2017). Gilgamesh locates this marine plant and returns to the shore, in order to have it consumed by an old man and thus ensure its rejuvenating power (see Figure I.1). Before proceeding with his experiment, Gilgamesh returns to bathe and the precious plant is stolen and consumed by a snake. According to the legend, since this event, snakes have been able to renew their skin, but humans have not. This inability to regenerate skin, and thus prevent aging, condemns Gilgamesh and all his fellow creatures to not know immortality. In this Sumerian epic, eternal youth is therefore associated with a marine plant. However, the notion of a marine plant is very general and applies to seaweed, eelgrass and seagrass. Eelgrass (*Zostera marina*, *Zostera japonica*) and Posidonia (*Posidonia oceanica*, *Posidonia australis*) are mainly found in the northern hemisphere (Atlantic Ocean, Mediterranean) or southern hemisphere (Pacific Ocean), but are absent from tropical zones. As the *Epic of Gilgamesh* is set in the Arabian Gulf, some contemporary authors assimilate the famous plant of immortality to a seaweed and not a sea grass.

In ancient Rome, Pliny the Elder referred to the medicinal use of seaweed in his encyclopedia *Naturalis historia*. In particular, he wrote that the external application of seaweed had a beneficial effect in the treatment of gout and pain. However, there was an early split between East and West over the use of seaweed. This was mainly due to the Romans' perception of

1

Algae in Traditional Medicine

1.1. Traditional Asian medicines

1.1.1. *Chinese medicine*

Traditional Chinese medicine, or TCM, is an ancient medicinal practice. It is one of the three great traditional medicines along with Galenism (Arab world) and Ayurveda (India). Its origins date back to 3,000 BCE (Nestler 2002). It is best known worldwide for the practice of acupuncture. However, this medicine is not limited to this particular technique and its use is based on a set of prescriptions, some of which can be likened to a lifestyle (see Table 1.1).

Traditional Chinese medicine is based on a set of initially oral traditions, originating in the community, family or religious traditions. Depending on the ethnic group, they may also be based on magical beliefs. Some of these traditions are rooted in Shamanism, Buddhism or Taoism. The diversity of the origins and practices employed has led ethnologist Paul Unschuld to write that "there is not a traditional Chinese medicine, but traditional Chinese medicines" (Unschuld 2018).

Initially based on oral tradition, traditional Chinese medicine has developed over time into a written practice. The oldest text, *Yi Jing* ("Book of Changes"), was written by the Chinese emperor Fuxi.

One of his successors, Emperor Shennong, was the pioneer of herbal medicine, and the source of the *Bencao*, or "Treatise on Medicinal Matters".

Finally, in the 3rd century BCE, the knowledge acquired through traditional Chinese medicine was compiled in a medical encyclopedia known as *Huangdi Nei Jing*, or “The Yellow Emperor’s Classics of Internal Medicine”. This work, which covers almost 2,000 years of medicinal practice, was written by Emperor Huang Di, better known as the “Yellow Emperor”.

Traditional Chinese medicine is based on a curative as well as preventive approach. This medicine is based on various precepts, such as nutrition, exercise and body manipulation (see Table 1.1).

Acupuncture	**Healthy nutrition**	**Consumption of medicinal plants***	**Therapeutic massages**	**Movement exercises (Qi Gong, Tai Chi)**
Common (preventive-curative)	Daily (preventive)	Quite frequent (curative)	Quite frequent (curative)	Daily (preventive)

* Including algae.

Table 1.1. *Precepts of traditional Chinese medicine for the prevention or treatment of disease (according to Nestler (2002))*

Today, traditional medicine coexists with modern Western-style medicine in Chinese hospitals. Algae, marine plants par excellence, are used in traditional Chinese medicine for both preventive (healthy food) and curative purposes.

Traditional Chinese medicine often uses the brown alga *Sargassum* in the treatment of various pathologies, as well as in the treatment of pain resulting from inflammatory mechanisms (Chatterji et al. 2010; Liu et al. 2018).

There are approximately 400 species in the genus *Sargassum*, and only a limited number of species are used in traditional Chinese medicine (see Table 1.2). Most of them are included in the Chinese pharmacopoeia and are associated with pain relief treatments or chronic pathologies, such as goiters, bronchitis or edemas (see Table 1.2).

Latin name	Chinese common name	Therapeutic applications
Sargassum pallidum ***Sargassum confusum*** ***Sargassum fusiforme***	Hai Zao	Treatment: – goiters – scrofula – testicular swelling and pain – edema – chronic bronchitis – angina pectoris – acute esophagitis – arteriosclerosis
Sargassum fulvellum ***Sargassum henslowianum*** ***Sargassum thunbergii*** ***Sargassum horneri***	Hai Qian	Treatment: – goiters – sore throat – cough – angina pectoris – edema – dysuria – boils – fever – pain
Sargassum siliquastrum ***Sargassum muticum*** ***Sargassum hemiphyllum*** ***Sargassum polycystum*** ***Sargassum vachellianum***	Species used in China, but not officially referenced under a common name	Same uses as for Hai Zao and Hai Qian

Table 1.2. *Examples of the use of* Sargassum *species (*Sargassum *spp.) in traditional Chinese medicine in the treatment of various pathologies (according to Liu et al. (2012))*

Sargassum algae, listed under the common name of “Hai Zao”, have been recognized in the Chinese pharmacopoeia since 1953. In the modern pharmacopoeia, in addition to the treatments listed in Table 1.2, sargassum is also recommended for the treatment of high blood pressure (Liu et al. 2012). According to the pharmacopoeia, the dose of seaweed should be between 6 and 12 g (expressed as dry matter). The method of administration is varied and depends on the nature of the ailment being treated. For dermal conditions, the product is administered as a powder spread topically over the lesion. However, for the treatment of certain internal dermal lesions, the product can be administered *per os* through the use of an extract. Algal decoction is one of the main ways in which an alga is administered in traditional therapeutic practice.

Algae can also be provided in the form of a food as an accompanying dish (sea vegetable). This is the case, for example, with the species *Sargassum fusiforme*, known as the “vegetable of longevity”. Traditional Chinese medicine, through this method of dietary administration of a drug, has appropriated the famous concept of “functional food”, little recognized in the Western world.

The species of sargassum known as “Hai Qiang” is part of folk medicine, especially in southeast China.

In addition to sargassum, other species belonging to the Chlorophyte (green algae) or Rhodophyte (red algae) phyla are also used in traditional Chinese medicine (see Table 1.3). These algae are recommended for the treatment of chronic conditions such as hypertension and hypercholesterolemia.

Chronic oral abscesses can also be treated by ingesting a seaweed decoction called “Long-shu-tsai”, which is prepared from the red alga *Gracilaria lemaneiformis* (Vonthron-Sénécheau 2016).

In traditional Chinese medicine, seaweed is also combined with other plants. They are used in complex preparations to achieve synergistic therapeutic effects. Sargassum is used in 226 herbal preparations.

Some, recognized by the Chinese pharmacopoeia, are highly complex preparations involving a multitude of ingredients (see Table 1.4). Such is the case with the medicinal preparation “Ru Jin Ling Ke” (乳疾灵颗粒), which combines more than a dozen ingredients that also include an animal organism (oyster or *Concha ostreae*).

Algal species	Therapeutic use
Ulva clathrata (green alga) *Gracilaria arcuata* (red alga) *Porphyra* spp. (red alga)	Treatment of goiters Treatment of inflammation of the lymphatic system
Chondrus ocellatus (red alga) *Gelidiella acerosa* (red alga) *Gelidium* spp. (red algae) *Gracilaria* spp. (red alga) *Hypnea* spp. (red alga)	Treatment of chronic constipation
Gracilaria verrucosa (red alga) *Ulva linza* (green algae)	Diuretic
Porphyra spp. (red alga) *Ulva* spp. (green alga)	Treatment of hypertension Treatment of hypercholesterolemia

Table 1.3. *Examples of green, red and brown algae used in traditional Chinese medicine for the treatment of chronic pathologies*

Another, less complex preparation combines brown algae of the *Laminaria* or *Sargassum* with a more limited number of terrestrial plants (7) and is also recognized by Chinese pharmacopoeia. This preparation, called "Xiao Ying Wu Hai Wan" (消瘿五海丸), is administered as an oral expectorant in pill form. Its therapeutic indications are associated with the treatment of goiters, tuberculosis of the lymph nodes or cancer[1].

1 See: http://www.tcmip.cn/ETCM/index.php/Home/Index/fj_details.html?id=3176.

Name of medicinal preparation	Composition	Therapeutic use	Mode of administration	Reference
Ru Ji Ling Ke Li (乳疾灵颗粒)	*Radix bupleuri* *Rhizoma cyperi* *Radix paeoniae rubra* *Salviae miltiorrhizae radix* *Salviae miltiorrhizae rhizoma* *Concha ostreae* *Citri reticulatae pericarpium* *Sargassum* *Laminaria* *Semen cuscutae* *Semen vaccariae* *Caulis spatholobi* *Herba epimedii*	Treatment: – liver stagnation syndrome – breast hyperplasia – pulmonary nodules	Oral (granules)	Chinese Pharmacopoeia Encyclopedia of Traditional Chinese Medicine

Table 1.4. *Example of a medicinal preparation incorporating the brown algae* Sargassum *and* Laminaria *(Ru Ji Ling Ke Li) (according to Liu et al. (2012), Chinese pharmacopoeia (see: http://www.tcmip.cn/ETCM/index.php/Home/Index/fj_details.html?id=2324))*

Algal species	Traditional use	Pharmacopoeia
Digenea simplex (red alga)	Antihelminthic	Chinese
Grateloupia livida (red alga)	Antihelminthic	Chinese
Corallina pilulifera (red alga)	Antihelminthic	Chinese
Gracilaria caudata (red alga)	Pest control	Chinese
Codium fragile (green alga)	Antihelminthic	Chinese
***Enteromorpha* spp.** (green alga)	Antihelminthic	Chinese
Udotea flabellum (green alga)	Antimicrobial	Chinese
Sargassum fluitans (brown alga)	Antihelminthic, antiprotozoal	Chinese
Sargassum thunbergii (brown alga)	Antimicrobial	Chinese
Sargassum horneri (brown alga)	Antimicrobial	Chinese

Table 1.5. *Examples of green, red and brown seaweed used as antiparasitic agents in traditional Chinese medicine (according to Vonthron-Sénécheau (2016))*

In traditional Chinese medicine, the use of seaweed is not restricted to the treatment of pain, goiters or certain chronic pathologies. Various species belonging to the red (Rhodophytes), green (Chlorophytes) or brown (Phaeophytes) algae groups are used in parasitic treatments of helminthic or protozoan origin (Vonthron-Sénécheau 2016). These uses are well known and listed in the Chinese pharmacopoeia (see Tables 1.4 and 1.5).

In Asia, other traditional medicines also use algae for therapeutic purposes. Traditional Japanese medicine is one such example.

1.1.2. *Japanese medicine*

Traditional Japanese medicine, or Kampo, is more recent than traditional Chinese medicine. It first appeared on the Japanese peninsula around the 5th century AD. It was introduced via Korea, which had itself received it from China. This medicinal practice is therefore derived from Chinese folk medicine. Kampo is a traditional medical approach used by over 80% of Japanese doctors (Motoo et al. 2011).

In Japanese clinical practice guidelines, 10% of prescriptions are associated with the use of Kampo. By comparison, only 7% of traditional Chinese medicine practices are listed in Chinese clinical practice guidelines.

In 2001, the Japanese government introduced the teaching of Kampo in medical studies. In this way, Japan has encouraged the fusion of two types of medicine: traditional and Western. This close association between the two types of medicine also exists in China but in a more limited way (see Figure 1.1). However, it is absent in Korea, where traditional medicine and Western medicine represent two separate branches of medicine (see Figure 1.1).

Kampo medicine is based on the use of therapeutic herbal formulas. The Japanese pharmacopoeia recognizes 165 herbal ingredients that can be used in Kampo preparations. The main manufacturer of Kampo medicines (the Tsumura firm) commonly uses licorice (*Glycyrrhizae radix*) and ginger rhizome (*Zingibiris siccatum rhizoma*) in its formulations. Seaweed is used less frequently, except in certain cases where the aim is to reinforce the effectiveness of Kampo medical practice.

In the case of digestive disorders or chronic pathologies relating to the digestive system, the Kampo approach is known as “Daikenchuto”. It is based on the use of plant preparations of various compositions.

The preparation consisting of *Zanthoxylum fructus* (Japanese pepper), *Zingibiris siccatum rhizoma* (ginger), *Ginseng radix* (ginseng) and maltose powder can be recommended to relieve the effects of colitis (Shi et al. 2022). The “Daikenchuto” approach is also recommended in the treatment of post-operative gastrointestinal ileus (Endo et al. 2017).

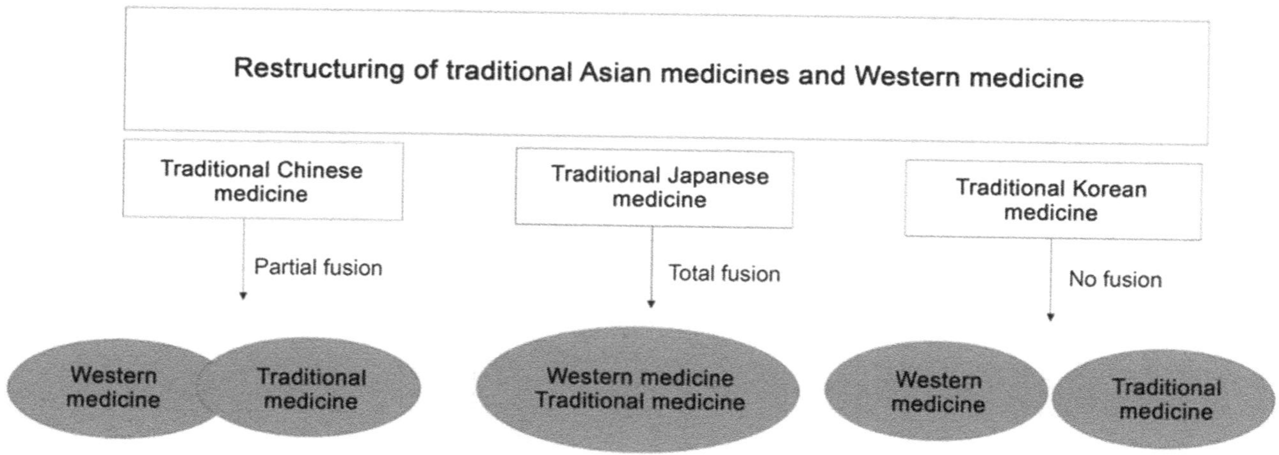

Figure 1.1. *Association or integration of traditional Asian medicines and Western medicine according to Motoo et al. (2011))*

In the case of stomach cancer, the use of Kampo medicine is recommended to reduce the pain and side effects of chemotherapy. The inclusion of algae in the basic formula of the "Daikenchuto" preparation is recommended during palliative care related to stomach cancer (Kobayashi 2016). The species concerned are mainly brown algae rich in fucoidanes and sulfated polysaccharides known for their immunostimulant and anticancer properties. These algae can be administered as extracts or as part of the patient's diet (see Table 1.6) (Kobayashi 2016). In the latter case, the term "Yakuzen", or "healthy medicinal cuisine", is used to describe the use of "vegetable seaweed" as a functional food.

Algal species	Common name
Laminaria japonica (*Saccharina japonica*) (brown alga)	Kombu
Undaria pinnatifida (brown alga)	Wakame
Hizikia fusiformis (brown alga)	Hijiki
Cladosiphon okamuranus (brown alga) *Nemacystus decipiens* (brown alga)	Okinawa – Mozuku Mozuku
Undaria pinnatifida (lower part of the thallus) (brown alga)	Mekabu
Gelidium elegans (red alga)	Tengusa
Sargassum horneri (brown alga)	Akamoku
Ulva lactuca (green alga)	Aosa
Porphyra yezoensis (red alga)	Nori

Table 1.6. *Algal species used in traditional Japanese Kempo medicine for the preparation of the enriched therapeutic preparation "Daikenchuto" or for healthy medicinal cooking (according to Kobayashi (2016))*

Apart from the classic field of Kampo medicine, certain species of algae are used in their own right in local traditional medicine practices. For example, in the Amami Islands, in the south of the Japanese archipelago (see Figure 1.2), certain red algae are used to produce deworming medicines.

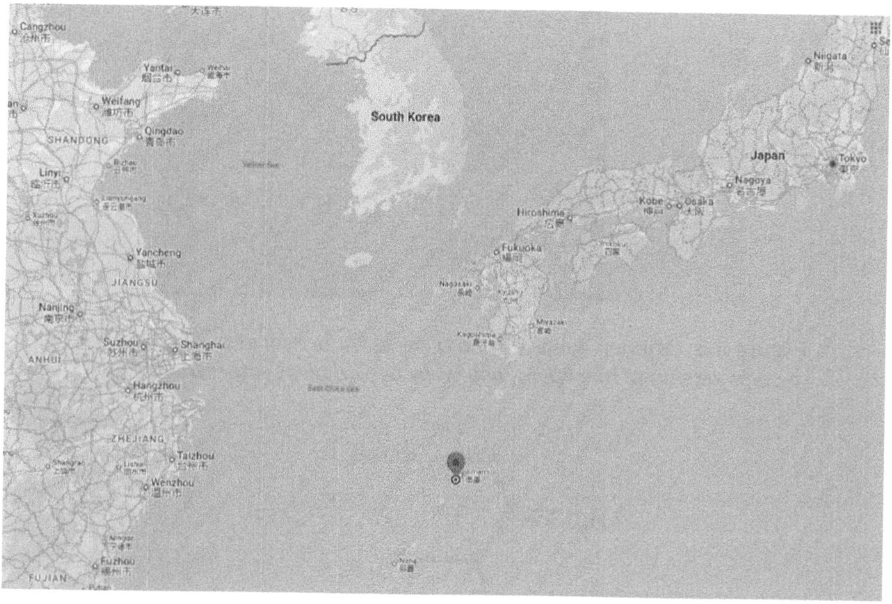

Figure 1.2. *Geographical location of the Amami Islands (according to Google Maps (2023)). For a color version of this figure, see www.iste.co.uk/fleurence/health.zip*

This is the case for *Digenea simplex* (see Figure 1.3) and *Chondria armata*, known as "Makuri" or "Hana-Yanagi", respectively (Arasaki and Arasaki 1985; Terada and Watanabe 2016).

In particular, these algae produce kainic (see Figure 1.4) and domoic acids, which are powerful antihelminthics (Maeno et al. 2019).

Algae from the family *Laminariacea* are also used in traditional Japanese medicine for their hypotensive effects. This activity is linked to the presence in these brown algae, and more specifically in the species *Laminaria angustata,* of a peptide called laminin (Girard et al. 1988). This hypotensive agent has also given rise to a drug known today under the generic name of Laminin (Arasaki and Arasaki 1985).

Figure 1.3. Digenea simplex *(photo credit © Lerissel et al. 2021). For a color version of this figure, see www.iste.co.uk/fleurence/health.zip*

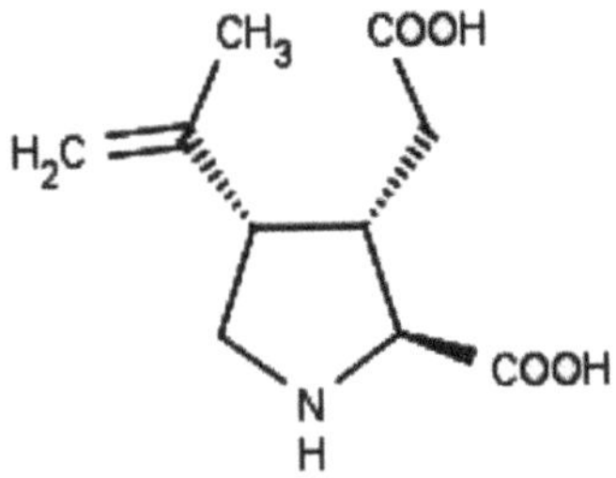

Figure 1.4. *Structure of kainic acid*

Kampo is also associated with the treatment of neurodegenerative pathologies such as dementia and Alzheimer's disease. The practice is based on a phytotherapeutic approach known as "Yokukansan". The associated preparation is essentially composed of the roots or rhizomes of terrestrial plants.

"Yokukansan" is recommended for reducing the severity of behavioral disorders associated with neurodegenerative diseases. In the case of dementia, it is recommended to alleviate certain symptoms, such as excitement, aggressiveness and hallucinations. Stemming from the Sino-Japanese tradition, this preparation has often been used in China to treat aggression in children.

Animal studies show the presence, in the preparation, of bioactive substances with neuroprotective and anti-stress effects (Czelusta et al. 2018). Such compounds also exist in brown algae. They are mainly polysaccharides (laminarinfucoidanes) (Pereira and Valado 2021). The addition of brown algae of the genus *Laminaria* (*Saccharina*) in the preparation "Yokukansan" may therefore be justified for the treatment of neurodegenerative diseases. Despite this, the scientific literature does not mention the association of algae with the basic "Yokukansan" preparation. It is likely that such a practice exists, but its application is surely local and must be based on an oral tradition not easily identifiable outside the regional context.

Figure 1.5. *Donguibogam Encyclopedia, on display at the National Museum of Korea (photo credit © Durrey J.)*[2]

1.1.3. *Korean medicine*

Korean traditional medicine is as old as Chinese traditional medicine. Its origins are generally estimated at 3,000 BCE. Like other traditional Asian medicines, it is based on the use of herbal beverages and the consumption of foods considered to be healthy. The practices of traditional Korean medicine have been recorded in an encyclopedia called *Donguibogam* (see Figure 1.5). This 25-volume medical encyclopedia was written in the 16th century by a physician at the court of King Seonjo of Joseon. This document, preserved in the National Library of Korea, is listed by UNESCO

2 See: https://commons.wikimedia.org/w/index.php?curid=47255719.

in the Memory of the World Register. It is the first medical work to be included in this prestigious register[3].

Figure 1.6. Codium fragile *(Cheong-kag), a green alga used in traditional Korean medicine (photo credit © Barbaroux O. 2016). For a color version of this figure, see www.iste.co.uk/fleurence/health.zip*

The links between traditional Chinese and Korean medicines are numerous, and certain prescriptions based on the use of algae to treat goiters in particular are shared between the two types of medicine. Traditional Korean medicine, however, presents a number of particularities. It specifically employs several species of green algae in extract form. The species *Codium fragile* and *Caulerpa lentillifera* (see Figures 1.6 and 1.7) are used to treat chronic or specific pathologies (see Table 1.7). *C. lentillifera*, commonly known as "Ba-da-po-do", is associated with the treatment of hypertension. The presence of hypotensive peptides recently characterized in this alga explains its empirical use in traditional Korean medicine (Joel et al. 2018).

The practices of traditional Korean medicine are largely based on those of traditional Chinese medicine. This is mainly due to the geographical, cultural and historical proximity between the two countries. Korea's adoption of the Chinese script (Kugyol) has made it possible to read the writings developed by traditional Chinese medicine and thus ensured their application in Korean medicine. Apart from this historical dimension, China and Korea share a

3 See: https://fr.unesco.org/courier/novembre-2009/donguibogam-book-precieux-medicine-coreenne.

common algal resource and similar practices in terms of both food and therapeutic uses.

Figure 1.7. Caulerpa lentillifera, *a green alga used in traditional Korean medicine (photo credit © Jpatokal 2009)*[4]*. For a color version of this figure, see www.iste.co.uk/fleurence/health.zip*

Algal species	Korean name	Use in traditional medicine
Capsosiphon fulvescens (green alga)	Mae-saeng-i	Treatment of stomach disorders Treatment of hangovers
Caulerpa lentillifera (green alga)	Ba-da-po-do	Treatment of hypertension Treatment of rheumatism Diabetes treatment Treatment of bacterial and fungal infections
Codium fragile (green alga)	Cheong-gak	Treatment of dysuria Treatment of edema Treatment of enterobiosis
Gloiopeltis tenax (red alga)	Pul-ga-sa-ri	Treatment of diarrhea Treatment of colitis

Table 1.7. *Examples of green and red algae used in traditional Korean medicine (according to Sanjewa et al. (2018))*

1.1.4. *Vietnamese medicine*

Like other Asian folk medicines, traditional Vietnamese medicine is inspired by traditional Chinese medicine. It is also based on the use of

4 See: https://commons.wikimedia.org/w/index.php?curid=6913523.

medicinal plants. Unlike the Chinese tradition, however, it does not rely on the use of complex mixtures. Instead, it relies on the use of simple preparations made from fresh or dried local plant ingredients. This medicine is also referred to as "southern herbal", in contrast to Chinese phytotherapeutic practice, which comes from the north.

In addition to ingestible preparations, traditional Vietnamese medicine also makes use of ointments and poultices.

Unlike the previous examples, seaweed is not widely used in this traditional medicine. This is mainly due to the recent nature of seaweed use in Vietnam. This resource has only been exploited by local populations for the past hundred years (Nang and Dinh 1998), and its use outside coastal areas is far from widespread. Despite this, the scientific literature mentions a few examples of the use of seaweed in Vietnamese medicine. These are mainly red algae (*Gracilaria* spp., *Kappaphycus cottonii*, *Caloglossa leprieurii*, and brown algae (*Sargassum* spp.) (Zemke-White and Ohno 1999). The fields of therapeutic application are not always well explained.

1.1.5. *Filipino medicine*

Algae are often used in traditional Filipino medicine, particularly in certain northern regions of the archipelago. Such is the case of the Ilocos Norte region, located to the far north of Manila (see Figure 1.8).

On its coastline, 34 species of algae are frequently used by the local population for the treatment of various pathologies, such as digestive, thyroid or respiratory disorders (see Table 1.8). These algae belong to the red, brown and green algae families.

The green alga *Caulerpa racemosa* (see Figure 1.9) is one of the species most frequently used under the local name of "Ar-arusip" (see Table 1.8). It is particularly recommended for the treatment of respiratory ailments, such as coughs and asthma (Dumilag and Javier 2022).

Other caulerps, such as *C. serrulata* or *C. sertularioides*, are also recommended in the treatment of these respiratory ailments (see Table 1.8). As in the case of traditional Japanese medicine, Filipino folk medicine recommends the use of the red algae *Digenea simplex* for its vermifuge properties.

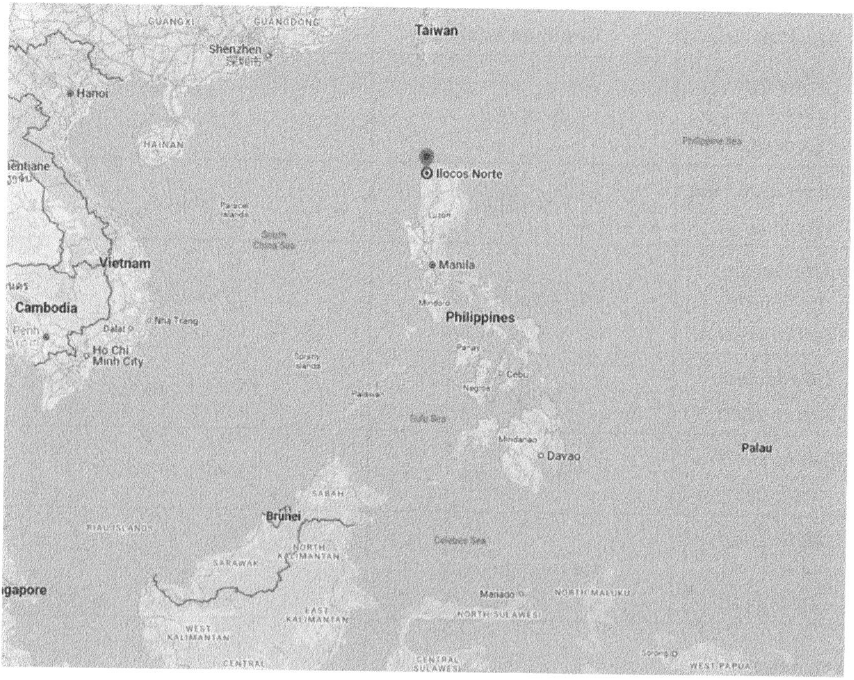

Figure 1.8. *Ilocos Norte region in the Philippine archipelago (from Google Maps (2023)). For a color version of this figure, see www.iste.co.uk/fleurence/health.zip*

Figure 1.9. *Caulerpa racemosa (photo credit © Hobgood N. 2007)*[5]*. For a color version of this figure, see www.iste.co.uk/fleurence/health.zip*

5 See: https://commons.wikimedia.org/w/index.php?curid=5704057.

Algal species	Common name	Therapeutic use
Caulerpa racemosa (green alga)	Ar-arusip	Cough Asthma
Codium arabicum (green alga)	Pok-poklo	Goiters
Halimeda incrassata (green alga)	Sal-salamu-gui	Goiters
Ulva lactuca (green alga)	Gam-gamet	Dermal infection
Digenea simplex (red alga)	Bodo-bodo	Parasitic infection
Halymenia durvillei (red alga)	Gayong-gay-ong	Indigestion
Gracilaria edulis (red alga)	Cao-caoy-an	Abdominal pain Antidiarrheal Asthma treatment
Palisada perforata (red alga)	Tar-tariptip	Antidiarrheal
Scinaia hormoides (red alga)	Garganatis	Sexual impotence
Padina minor (brown alga)	Lap-lapayag	Antidiarrheal
Sargassum aquifolium (brown alga)	Aragan	Indigestion

Table 1.8. *Examples of algal species used by traditional Filipino medicine (according to Dumilag and Javier (2022))*

Gastrointestinal disorders, such as indigestion or abdominal pain, can also be treated using various species belonging to the genera *Sargassum* or

Gracilaria (see Table 1.8). Symptoms of digestive disorders, such as diarrhea, are also treated with algae. The species concerned by this particular application belong to the genus *Padina* (see Figure 1.10). Traditional Filipino medicine also uses algae to treat thyroid dysfunction, the cause of goiters. For this therapeutic application, it relies mainly on green algae of the genera *Codium* and *Halimeda* (Dumilag and Javier 2022). More anecdotally, this medicine sometimes recommends the use of certain species to treat sexual impotence. These are mainly red algae belonging to the species *Phycocalidia acanthophora* and *Scinaia hormoides*.

Figure 1.10. Padina *sp. (Lap-lapayag), a brown alga used in traditional Filipino medicine for the treatment of diarrhea (photo credit © Fleurence J. 2023). For a color version of this figure, see www.iste.co.uk/fleurence/health.zip*

Seaweed can be administered in a variety of ways. For *per os* use, decoction remains the most widespread method. Next come infusions, alcoholic extracts and syrups. For topical use, an alga is applied to lesions in the form of creams or poultices.

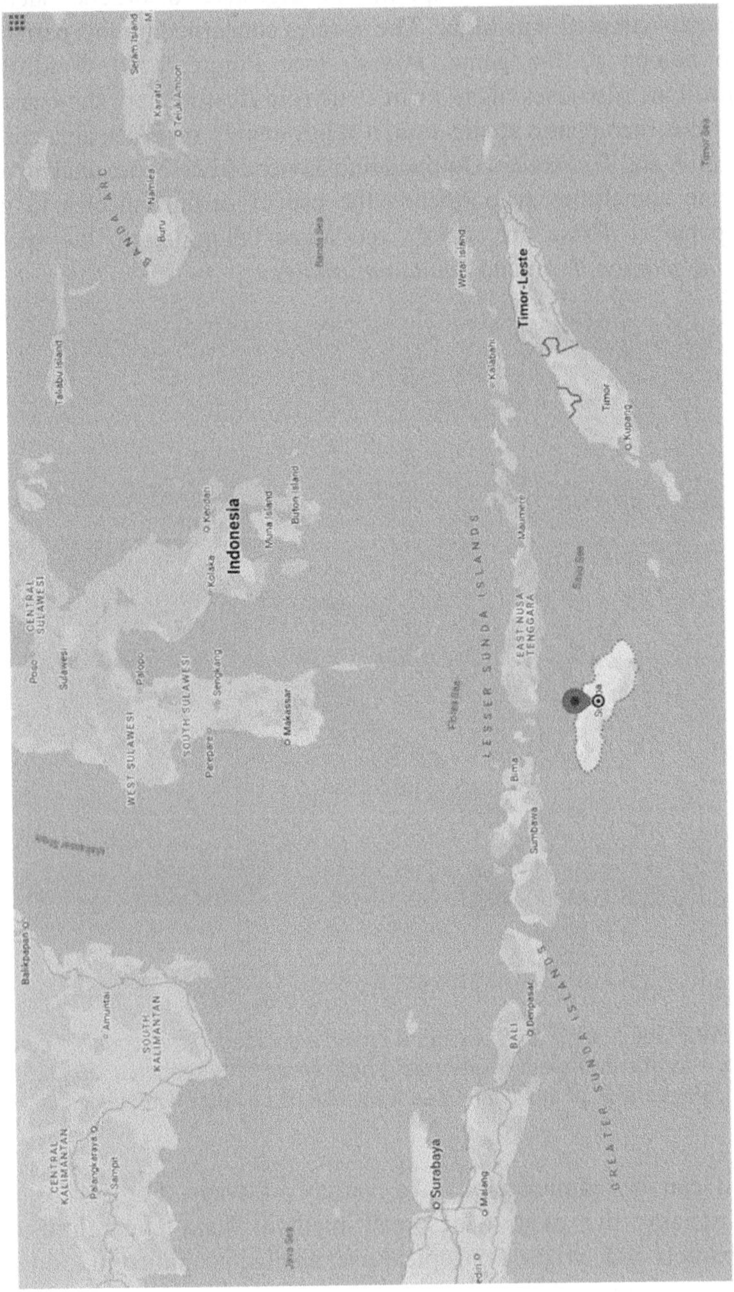

Figure 1.11. *Location of the island of Sumba in the Indonesian archipelago (from Google Maps (2023)). For a color version of this figure, see www.iste.co.uk/fleurence/health.zip*

Folk medicine in the Philippines differs from other traditional Asian medicines by the use of a large number of algal species and by the diversity of pathologies or ailments treated. It is based on local, ancient practices that are now well documented in the recent literature. Unlike traditional Chinese or Korean medicine, it is not based on a thousand-year-old medical encyclopedia, but rather on the oral traditions of coastal populations. The familiarity of these populations with the use of algae as a food resource probably explains the interest they have developed for algae in folk medicine.

1.1.6. *Indonesian medicine*

Indonesian folk medicine includes algae in its therapeutic practices. These practices are often localized to certain islands in the archipelago. This is the case of the island of Sumba, located between the main island of Java and Timor (see Figure 1.11).

An ethnobotanical study carried out on the island reveals that 38 species of algae belonging to 18 different genera are used in local traditional medicine (Anggadiredja 2009). Of the algal species listed as "medicinal plants", the majority are brown algae (see Figure 1.12).

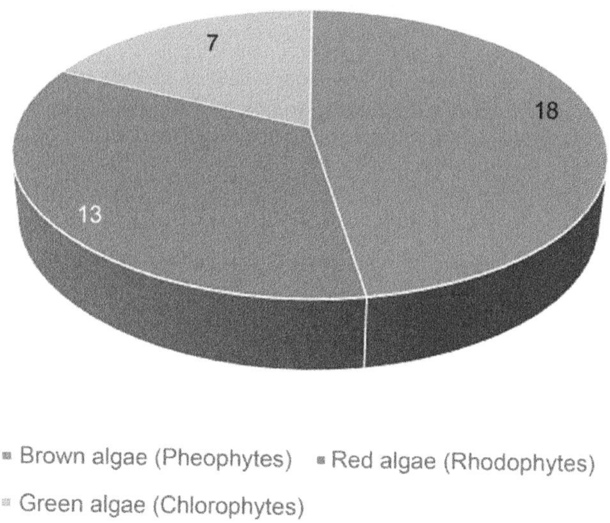

Figure 1.12. *Distribution of the different taxonomic groups of algae used in traditional medicine on the island of Sumba (Indonesia) from Anggadiredja (2009). For a color version of this figure, see www.iste.co.uk/fleurence/health.zip*

The main therapeutic application of algae on the island of Sumba is the treatment of goiters. For this type of treatment, brown algae are still the most popular (see Figure 1.13). This is partly due to the high iodine content of Pheophytes, compared with species belonging to the other two groups (personal communication, Fleurence J.).

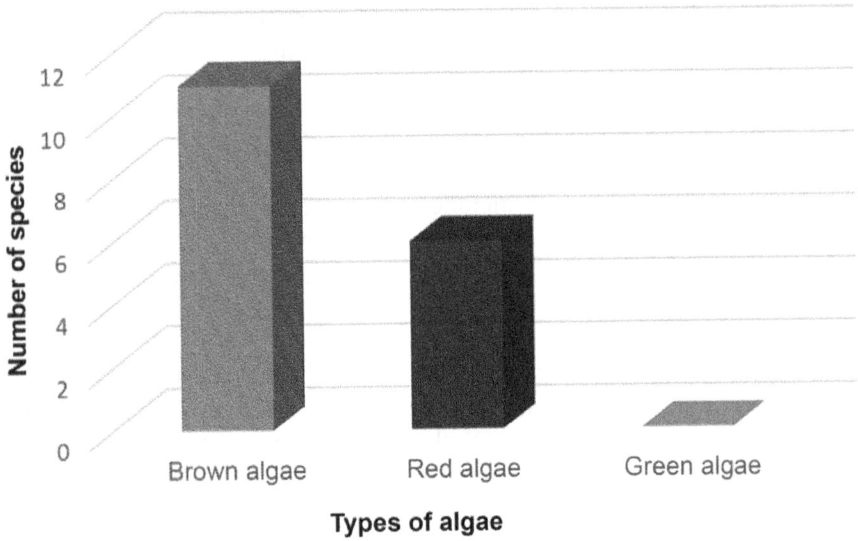

Figure 1.13. *Distribution by the algal group of species used for the treatment of thyroid dysfunctions by traditional Indonesian medicine (goiters) (according to Anggadiredja (2009))*

As with other traditional Asian medicines, Indonesian folk medicine uses algae to treat a wide range of respiratory, digestive and dermal pathologies (see Table 1.9).

The species *Codium arabicum* is used in traditional Indonesian and Filipino medicine, but for very different therapeutic applications (goiters or vermifuge) (see Tables 1.8 and 1.9).

This highlights the local nature of traditional medicine in the Philippines and Indonesia. This is in stark contrast to Chinese traditional medicine, which has disseminated and fed into Korean and Japanese traditional medicines.

Algal species	Therapeutic use
Codium arabicum (green alga)	Deworming Treatment of urinary diseases
Chaetomorpha linum (green alga)	Cough suppressant Antipyretic Antiseptic
Ulva lactuca (green alga)	Antipyretic Nasal antihemorrhagic Treatment of urinary diseases
Eucheuma serra (red alga)	Treatment of goiters Treatment of scrofula Stomach upset treatment
Gracilaria arcuata (red alga)	Treatment of goiters Treatment of scrofula Stomach upset treatment
Gracilaria gigas (red alga)	Treatment of goiter Treatment of scrofula Treatment of urinary diseases
Laurencia obtusa (red alga)	Antiseptic Stomach upset treatment
Hypnea musciformis (red alga)	Antipyretic Antiseptic
Ascophyllum sp. (brown alga)	Deworming Laxative Treatment of goiters
Sargassum ilicifolium (brown alga)	Treatment of goiters Treatment of scrofula
Padina japonica (brown alga)	Treatment of goiters Treatment of scrofula

Table 1.9. *Examples of algal species used by traditional Indonesian medicine (according to Anggadiredja (2009))*

1.2. Other traditional medicines

1.2.1. *African medicine*

Like all folk medicines, traditional African medicine relies on the use of plants in its therapeutic practice. It draws not only on herbalism, but also on spirituality. Some of these uses are now well documented in ethnopharmaceutical publications (Van Wyck 2008). Some of this work has focused on the medicinal practices of two African communities, the Koikhoi and San ethnic groups. The first, also known as the Khoi, is a nomadic, pastoral population living in South Africa. The San ethnic group is a group of hunter-gatherer populations also living in southern Africa.

Figure 1.14. Ecklonia maxima *(photo credit © Derekkeats 2010, uploaded by JoJan)*[6]*. For a color version of this figure, see www.iste.co.uk/fleurence/health.zip*

These ethnic groups practice a traditional medicine based on the use of a large number of plant species, mainly from the terrestrial environment. However, their medicinal practice also includes the use of algae. The most widely used species is the brown alga *Ecklonia maxima* (see Figure 1.14), better known as "sea bamboo" and very common on the coasts of South Africa and Namibia.

This alga is recommended for treating iodine deficiency, glandular swelling and syphilis. It is administered in the form of hot poultices or infusions.

6 See: https://commons.wikimedia.org/w/index.php?curid=12225732.

1.2.2. *European medicine*

In Europe, some countries have incorporated algae into their traditional medicinal practices. This is the case in Bulgaria, a country on the Balkan Peninsula with a long coastline on the Black Sea. On this coast, the main species is the brown alga *Cystoseira barbata*. In traditional Bulgarian medicine, this species is used in the form of decoctions and for the treatment of chronic pathologies such as hypertension and diabetes (Stoyneva-Gärtner and Uzunov 2017). In addition to this species, which is common on the Black Sea coast, traditional Bulgarian medicine also makes use of other algal species that are rarely or never found on the country's coastline. These include *Fucus vesiculosus* and *Chondrus crispus*, algae usually found on Atlantic coasts. Like *Cystoseira barbata*, these species are used for the treatment of various chronic or transient pathologies (see Table 1.10).

Algal species	Form of administration	Pathology treated
Cystoseira barbata (brown alga)	Decoctions	Hypertension Obesity Diabetes Atherosclerosis
Chondrus crispus (red alga)	Decoctions	Chronic bronchitis Cough Graying hair Pyrosis (acid reflux) Hyperthyroidism
Fucus vesiculosus (brown alga)	Baths Massages	Rheumatism Tendonitis Muscular tetany Poliomyelitis

Table 1.10. *Examples of treatments using algae in Bulgarian traditional medicine (according to Stoyneva-Gärtner and Uzunov (2017))*

Apart from Bulgaria, other European countries have used algae in their traditional therapeutic practices. These are often brown algae belonging to the genera *Laminaria* or *Fucus*. Within the genus *Fucus*, the species *Fucus serratus* (see Figure 1.15) and *Fucus vesiculosus* (see Figure 1.16) are the main fucals used in traditional European medicine. These two species are still listed as traditional medicinal plants in the French pharmacopoeia[7].

Figure 1.15. Fucus serratus *(photo credit © Fleurence J. 2023). For a color version of this figure, see www.iste.co.uk/fleurence/health.zip*

For traditional homeopathic preparations, however, only the species *F. vesiculosus* is recommended by the same pharmacopoeia[8].

In the United Kingdom, Ireland and France, algae were often recommended as a dietary supplement for sick children and the elderly (Pérez-Lloréns et al. 2023). A poster by the famous Italian illustrator

7 See: https://ansm.sante.fr/uploads/2023/01/10/liste-a-des-plantes-medicinales-utilisees-traditionally-january-2023-v2.pdf.

8 See: https://ansm.sante.fr/uploads/2020/10/22/fucus-vesiculosus-pph.pdf.

Leonetto Cappiello (1875–1942) perfectly sums up the recommendations of the time (see Figure 1.17). In particular, it extols the benefits of a marine food cure based on the consumption of a super-food, or "Maraliment", made from algae. But algae were also incorporated into traditional therapeutic practice, to treat a variety of chronic or temporary ailments (see Table 1.10).

Figure 1.16. Fucus vesiculosus, *a brown alga used in traditional homeopathic preparations (photo credit © Fleurence J. 2023). For a color version of this figure, see www.iste.co.uk/fleurence/health.zip*

In addition to brown algae, traditional European medicine also involves other algae, such as the Rhodophyte *Palmaria palmata*, which is commonly known as dulse or dillisk. Found on Europe's Atlantic coasts, this alga was traditionally used in Ireland and Scotland for its purifying, antipyretic and anti-scorbus properties. A Scottish saying went: "He who eats the Dulse of

Guerdie and drinks from the well of Kildinguie will escape all diseases except the Black Death".

In European Mediterranean countries, the red alga *Corallina officinalis*, or Coralline officinale, was also used in traditional medicine. Taken internally in the form of decoctions, infusions or syrups, this alga was mainly used as a vermifuge (Pérez-Llorens et al. 2023).

Figure 1.17. *Poster promoting the benefits of the Maraliment seaweed cure (according to Leonetto Cappiello (1875–1942)). For a color version of this figure, see www.iste.co.uk/fleurence/health.zip*

In France, a laxative based on agar or agar-agar was also recommended to restore proper intestinal function. This product, marketed at the beginning of the 20th century under the name Jubol, met with considerable popular success, not least thanks to numerous advertising campaigns for the time (see Figure 1.18). In addition to its functional and laxative properties, the drug was also recommended for the treatment of hemorrhoids and migraines. During World War I, Jubol was often recommended to French soldiers in the trenches, to combat constipation and hemorrhoids caused by precarious, stressful living conditions. In 1916, the French newspaper *Le Petit Provençal* published an advertisement in which Jubol was equated with the French soldier in the trenches dislodging the related German soldier, a microbe, who must be driven away (see Figure 1.19). This was one of the rare occasions when a traditional algae-based pharmaceutical product was associated with an act of war propaganda.

Algal species	Therapeutic application	Country
Laminaria spp. (brown alga)	Treatment of goiters	United Kingdom Ireland France
Laminaria stipes **(L. digitata, L. clouston)** (brown alga)	Surgery	Ireland
Fucus spp. (brown alga)	Treatment of goiters Treatment of scrofula	United Kingdom Ireland France
Alaria esculenta (brown alga)	Treatment of pica disease Stomach fortification Restoring normal appetite	Denmark (Faroe Islands)
Chondrus crispus (red alga)	Treatment of bronchitis Cough treatment Treatment of dysentery Treatment of diarrhea Treatment of scrofula	Ireland
Palmaria palmata (red alga)	Scurvy treatment Blood purification treatment Antipyretic	Scotland (Orkney Islands) Ireland

Table 1.11. *Examples of the use of algae in traditional Northern European medicine (according to Sauvageau 1920; Pérez-Llorens et al. 2023)*

Jubol, a traditional French medicine, has now been replaced by Gaviscon, a pharmaceutical preparation composed mainly of sodium alginate (47.5% of content). This contemporary medicine, based on phycocolloids, is mainly used in the treatment of acid reflux.

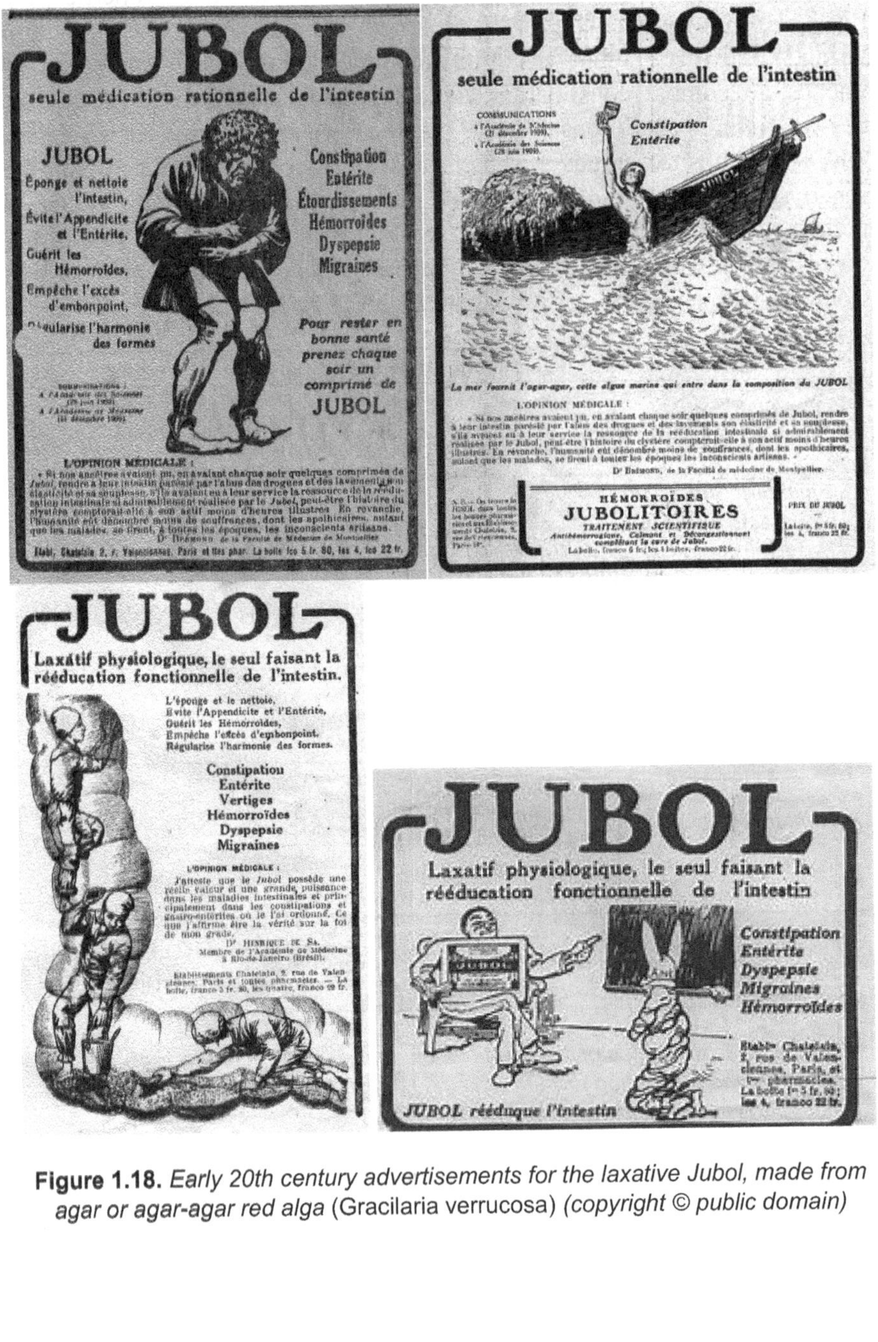

Figure 1.18. *Early 20th century advertisements for the laxative Jubol, made from agar or agar-agar red alga* (Gracilaria verrucosa) *(copyright © public domain)*

Figure 1.19. *War propaganda advertisement in* Le Petit Provençal *(1916): the qualities of the agar-based medicine Jubol (copyright © public domain)*

1.2.3. *Medicine on the American continent*

Type	Botanical group
Sargassum	Brown alga
Macrocystis	Brown alga
Durvillaea	Brown alga
Gigartina	Red alga
Mazzaella	Red alga
Sarcothalia	Red alga

Table 1.12. *Main seaweed genera used by prehistoric populations on the southern Chilean coast for food and medicinal purposes (according to Pérez-Lloréns et al. (2023))*

Evidence of the use of algae as a food and medicinal plant has been found at the Monte Verde archaeological site in southern Chile. These traces, which date back to around 14,000 BCE, demonstrate the role played by algae in the daily lives of populations settled along the southern coasts of the Pacific Ocean (Dillehay et al. 2008). The species concerned belong to different genera, mainly associated with the groups of brown algae (Phaeophytes) or red algae (Rhodophytes) (see Table 1.12).

In his 1856 book *Elementos de farmacia applicada a la medicina*, Professor Vicente Bustillos cites the use of algae as a remedy for lung ailments (gels for poultices), diarrhea, chronic dysentery, scrofula and oral infections (Bustillos 1856).

Apart from coastal populations, the inhabitants of central Chile, and in particular the Mapuche Indians, have also used algae in their medicinal practices (Bustillos 1856; Pérez-Lloréns et al. 2023). The species used, *Durvillaea antartica*, known as "cochayuyo", has been used for the treatment of mumps and for making poultices.

This traditional product is still marketed in Chile under its original name (see Figure 1.20).

Figure 1.20. *Bales of cochayuyo about to be sold at a market in Chile (photo credit © drapeau picard A.P. 2012)*[9]. *For a color version of this figure, see www.iste.co.uk/fleurence/health.zip*

9 See: https://commons.wikimedia.org/w/index.php?curid=20160956.

1.2.4. *Polynesian medicine*

Polynesia, or "many islands" according to Greek etymology, comprises a large number of archipelagos in the Pacific Ocean. Algae are not widely used as food or medicine in the islands of French Polynesia. It is absent in Tahiti, but present in the Marquesas Islands (Conte and Payri 2002). In the Marquisian archipelago, algae are mainly consumed as a sea vegetable under the generic name of "imu". These are mainly six species belonging to the genera *Enteromorpha*, *Cladophora*, *Ulva*, *Caulerpa* and *Chnoospora*. However, there are no written examples of the use of these algae in traditional Marquesan medicine.

The situation is different in the Hawaiian archipelago, where algae are not only traditionally consumed, but also incorporated into folk medicine. More than 60 species are used by Hawaiians for these two applications (McDermid et al. 2019). These algae are known generically as "limu". However, less than half are identified by the scientific binominal nomenclature, i.e. their Latin name. Many species of limu are used in traditional Hawaiian medicine. They are used to treat a wide range of ailments, including minor wounds, asthma, oral thrush, miscarriages and dietary problems. More surprisingly, algae are also associated with the treatment of psychological ailments. This therapeutic contribution is achieved through traditional rituals and ceremonies.

A case in point is the ceremony of forgiveness (ho'oponopono), which aims to resolve personal conflicts. During this ancient practice, each participant consumes a blade of *Sargassum aquifolium,* an alga commonly known as "limu kala". Consumption of the blade during the Indian ceremonial marks the end of the ritual and the release of interpersonal tensions. Limu kala is also used in healing or purification rituals for those affected by psychological suffering following bereavement.

In Hawaii, algae play an important role in popular culture. They are an integral part of traditional dietary and therapeutic practices. They are also part of the legends of this archipelago, and certain species, such as Ulva, qualified as "limu palahalaha" (see Figure 1.21), enjoy sacred status.

Figure 1.21. *Green alga* Ulva *sp. or* limu palahalaha, *considered sacred in traditional Hawaiian culture (photo credit © Fleurence J. 2019). For a color version of this figure, see www.iste.co.uk/fleurence/health.zip*

2

Activities of Therapeutic Interest

2.1. Antioxidant activities

The presence of antioxidant activities in algal extracts has often been described in the scientific literature. The extracts concerned are obtained by aqueous extraction or using organic solvents (ethanol, methanol, ethyl acetate, dichloromethane, etc.). This implies that the compounds carrying the activity are hydrophilic or hydrophobic in nature. The antioxidant activities of algal extracts have been characterized in numerous species belonging to the three major botanical groups (Pheophytes, Rhodophytes, Chlorophytes). The activity detected in these extracts is sometimes associated with an antimicrobial action. This latter property of therapeutic interest will be developed later (see section 2.2).

2.1.1. *Methodology*

Evidence of antioxidant activity is based on the evaluation of in vitro lipid peroxidation inhibitory activity in algal extracts. Lipid peroxides can be produced enzymatically (lipoxygenase) or by auto-oxidation (see Figure 2.1).

The antioxidant activity of the extract can be assessed by measuring lipoxygenase inhibitory activity or by determining the trapping activity of the free radical 2,2-diphenyl-1-picrylhydrazyl (DPPH). The two methods, although effective, can give different results in terms of estimating the antioxidant power of extracts (see Figure 2.2). This is the case of the ethanolic extract of *Sargassum siliquastrum*, whose lipoxygenase inhibitory activity is estimated at 68% and DPPH trapping action at approximately 90%.

Step 1 or initiation of oxidation	$LH \longrightarrow L^{\bullet}$
Step 2 or propagation of oxidation	$L^{\bullet} + O_2 \longrightarrow LOO^{\bullet}$
	$LOO^{\bullet} + LH \longrightarrow LOOH + L^{\bullet}$
Step 3 or end of peroxidation	$LOO^{\bullet} + LOO^{\bullet} \longrightarrow$ Non radical products

Figure 2.1. *Lipid auto-oxidation reactions by radical mechanism (according to Matsukawa et al. (1997)) (LH: reduced lipid, L.: lipid in radical form, LOO: radical peroxidized lipid, LOOH: hydroperoxidized lipid)*

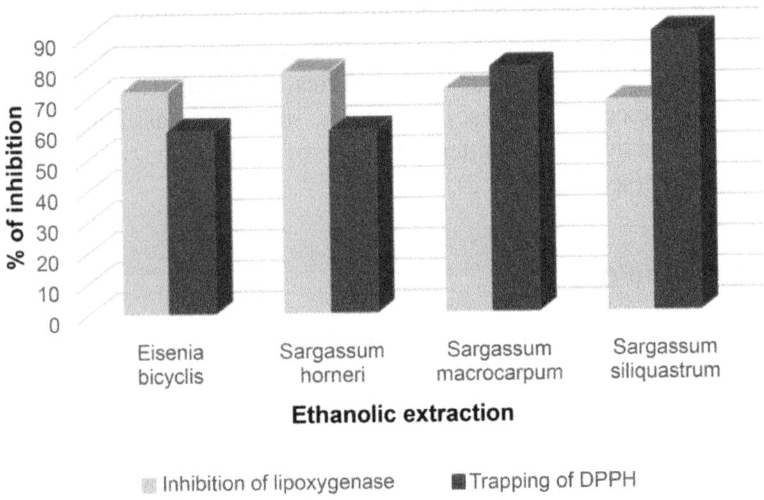

Figure 2.2. *Comparison between methods for assessing (lipoxygenase inhibition, DPPH trapping) of the oxidizing power of ethanolic extracts of brown algae (according to Matsukawa et al. (1997))*

2.1.2. *Active algae extracts*

2.1.2.1. *Aqueous extracts*

Antioxidant activity can be associated with fractions obtained by aqueous extraction. This shows that this type of activity can be carried out by hydrophilic molecules. However, the antioxidant activity of aqueous extracts is often lower than that reported for ethanolic or methanolic extracts. The latter solvents offer a broader extraction spectrum, incorporating polar molecules, i.e. relatively hydrophilic compounds, as well as amphiphilic compounds such as glycolipids or phospholipids.

This is particularly the case for extracts of the brown alga *Sargassum horneri* described in the literature for the high lipoxygenase inhibitory activity of its ethanolic extract compared with the aqueous extract (see Figure 2.3) (Matsukawa et al. 1997).

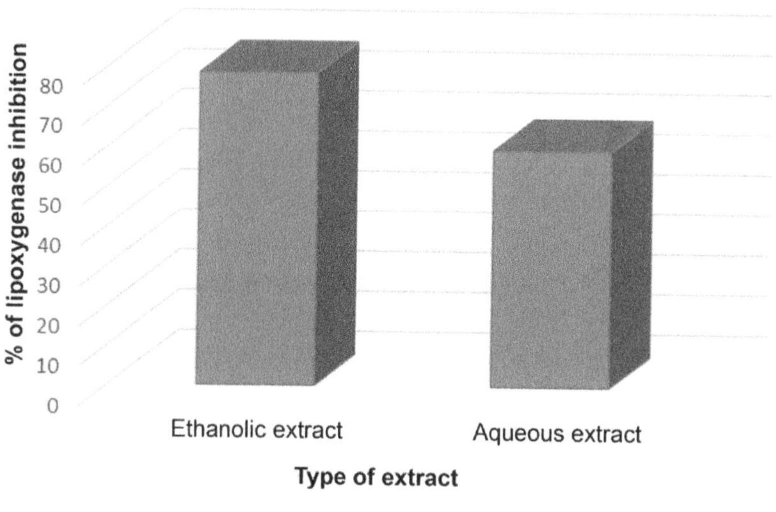

Figure 2.3. *Comparative efficacy of ethanolic and aqueous extracts of* Sargassum horneri *in lipoxygenase inhibition (according to Matsukawa et al. (1997))*

2.1.2.2. *Mixed extracts*

2.1.2.2.1. Water/organic solvent extracts

Combined water–methanol extraction can also be used to obtain an extract with antioxidant activity. This type of mixed extraction has been carried out on algae commonly consumed as sea vegetables (*Fucus*, *Laminaria*, *Undaria*, *Porphyra*) (Jimenez-Escrig et al. 2001). In this study, the antioxidant activity of algal extracts is assessed by their DPPH trapping activity. Free radical capture is assessed in relation to the quantity of algae required to neutralize 50% of the radicals present in the medium. The lower the quantity of algae needed to achieve this 50% inhibition rate of radical species, the higher the antioxidant activity of the extract. The highest antioxidant activity is found in *Fucus* (see Figure 2.4). This activity can be classified in descending order according to the following species: *Fucus* > *Laminaria* > *Undaria* > *Porphyra*.

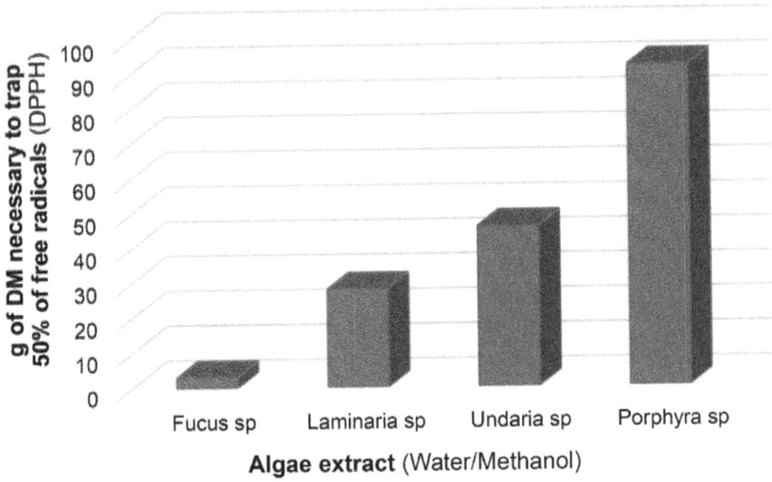

Figure 2.4. *Impact of algae concentration on the trapping of 50% DPPH-type free radicals (according to Jimenez-Escrig et al. (2001))*

2.1.2.2.2. Organic solvent/organic solvent extracts

Extraction using a mixture of polar and apolar organic solvents is also a very common process. The most frequently used combination is a chloroform/methanol mixture. This type of combination is used to obtain organic extracts of brown algae belonging to the genera *Laminaria*, *Himanthalia*, *Fucus* and *Ascophyllum* (Le Tutour et al. 1998). These algae, known in Europe under the generic name of wrack, are used in particular as edible algae. The extracts obtained show inhibitory activity against methyl linoleate oxidation. This antioxidant activity is synergistic with that of vitamin E. Paradoxically, for *Laminaria digitata*, this activity seems to depend on the harvesting period (winter, spring). Indeed, the extract obtained from algae harvested in winter shows a negative synergy with vitamin E in protecting against oxidation. This suggests the existence in *L. digitata* of pro-oxidant compounds present in winter and absent in spring. This study highlights the existence of a seasonal variation in antioxidant activity and the presence of pro-oxidant substances that are also dependent on the seasonal factor. However, this latter aspect is rarely described.

2.1.2.3. *Organic extracts*

Organic extracts are mainly obtained from an initial extraction in the presence of methanol. This methanol extract may or may not be fractionated using solvents of decreasing polarity, such as butanol, ethyl acetate, dichloromethane or petroleum ether.

The methanolic crude extract of the brown alga *Sargassum marginatum* shows significant antioxidant activity, measured by the DPPH method. This activity depends on the quantity of algae contained in the crude fraction (see Figure 2.5). This dose effect has also been reported with other brown algae, such as *Padina tetrastomatica* and *Turbinaria conoides* (Chandini et al. 2008).

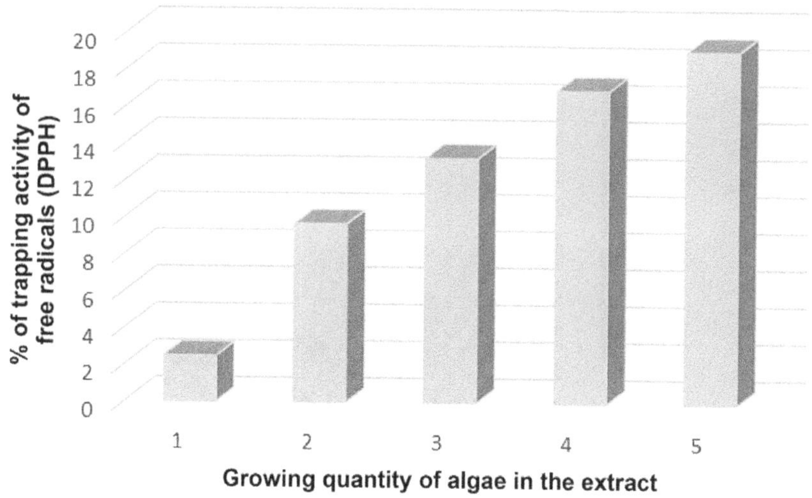

Figure 2.5. *Effect of an increasing amount of algae on the antioxidant activity of a crude methanolic extract of* Sargassum marginatum *(according to Chandini et al. (2008)) (1: 150 µg, 2: 400 µg, 3: 600 µg, 4: 800 µg, 5: 1,000 µg)*

In many studies, methanol appears to be the most efficient organic solvent in terms of the algal dry matter extraction yield compared to other solvents (see Figure 2.6).

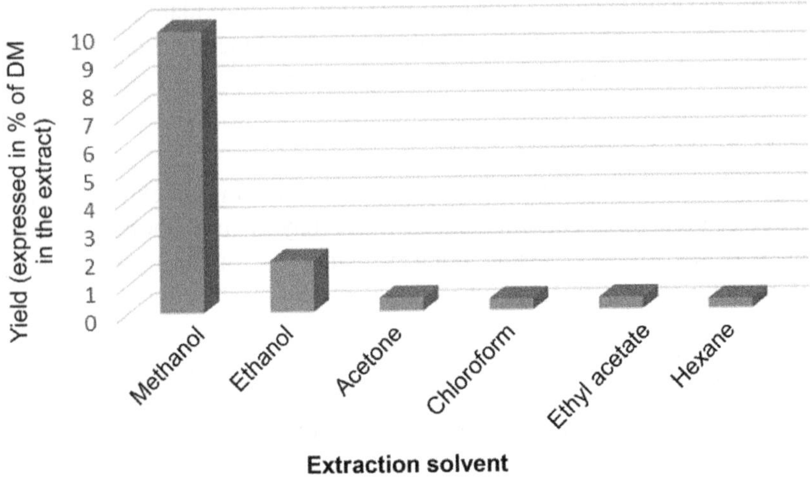

Figure 2.6. *Influence of the extraction solvent type on the extraction yield of dry matter of* Eisenia bicyclis *(arame) (according to Airanthi et al. (2011))*

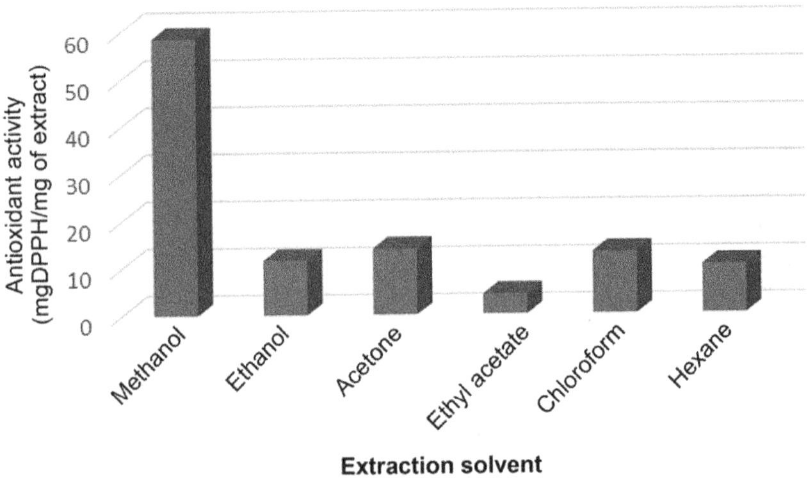

Figure 2.7. *Influence of the type of extraction solvent on the antioxidant activity of the edible alga* Eisenia bicyclis *(arame) (according to Airanthi et al. (2011))*

Irrespective of this, the highest antioxidant activity is often associated with methanolic extracts. This type of finding applies to many species, most notably *Eisenia bicyclis*, commonly consumed in Japan under the name arame (Airanthi et al. 2011). For this species, the highest antioxidant activity is indeed observed with the methanolic extract (see Figure 2.7). In addition to the extraction solvent aspect, antioxidant activity can differ according to the species considered (see Figure 2.8). This is particularly the case with certain algae commonly consumed as sea vegetables in Japan (see Table 2.1).

Species	Common name
Eisenia bicyclis	Arame
Kjellmaniella crassifolia	Gagome
Alaria crassifolia	Chigaiso
Sargassum horneri	Akamoku
Cystoseira hakodatensis	Uganomoku

Table 2.1. *Examples of brown algae consumed in Japan and tested for antioxidant activity in their methanolic extracts*

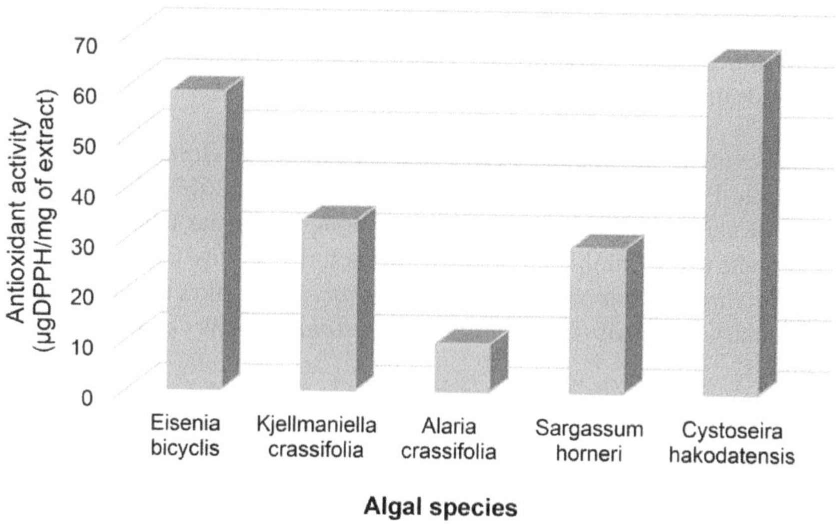

Figure 2.8. *Antioxidant activity of methanolic extracts obtained from different species of brown algae consumed as sea vegetables in Japan (according to Airanthi et al. (2011))*

2.2. Antibacterial activity

Numerous algae extracts have been described in the literature for their antibacterial activities. As before, these are aqueous extracts, organic extracts or solvent mixtures. The search for antibiotic activity is often combined with the characterization of antioxidant activity in the extracts concerned. The algae studied generally belong to species frequently used in human food. This type of research is therefore often associated with a global approach to the use of algae as food or active food ingredients.

The disk diffusion method is the most commonly used approach for screening antibacterial activity (see Figure 2.9). Its principle and implementation are straightforward.

First, the extract to be tested is applied to a sterile disk. In a second step, the impregnated disk is placed on a bacterial mat. The area where the bacterial colonies have not been able to grow is measured. The inhibition diameter thus determined characterizes the presence and intensity of antibacterial activity. The larger the diameter, the greater the antibiotic activity. This method provides mainly qualitative information and is ideally suited to screening studies.

Most antibacterial activities are reported for organic extracts, particularly ethanolic or methanolic extracts.

Algal extracts are mainly active against Gram+ bacteria (Hellio et al. 2000; Freile-Pelegrin and Morales 2004; Srikong et al. 2017). The ethanolic fraction of certain species of green algae belonging to the Ulvaceae family has been shown to inhibit the growth of bacterial strains belonging to the genera *Bacillus*, *Streptococcus* and *Staphylococcus* (Hellio et al. 2000). This is particularly true of *Enteromorpha intestinalis,* now known as *Ulva intestinalis*.

The latter's extract has proved particularly active in inhibiting the growth of *Bacillus megaterium* and a mixture of Gram+ bacteria (see Figure 2.10). However, it has no effect on the inhibition of Gram– forming bacteria such as *Serratia marcescens*.

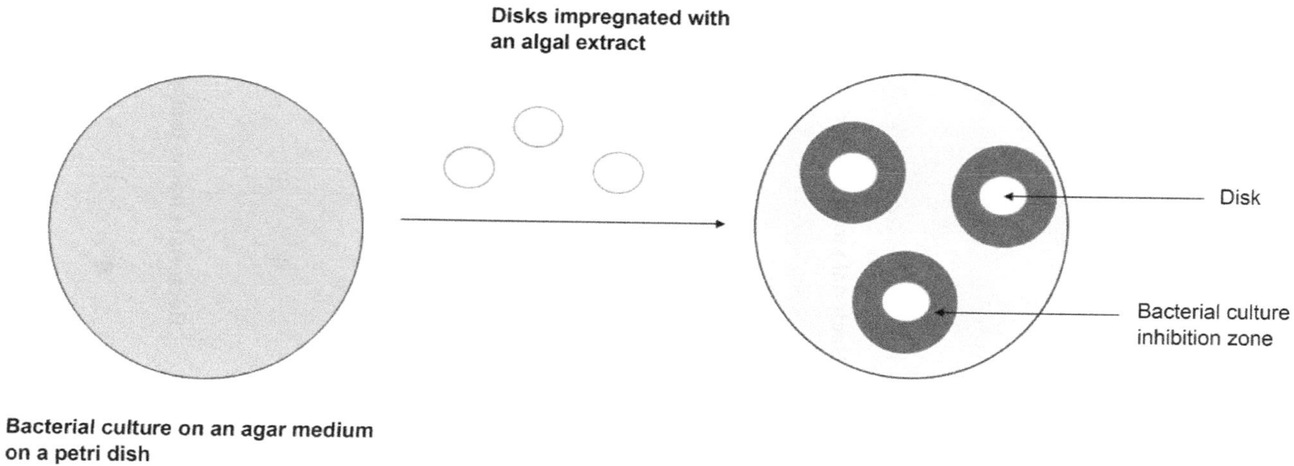

Figure 2.9. *Disk diffusion method used for screening antibacterial activities in algal extracts*

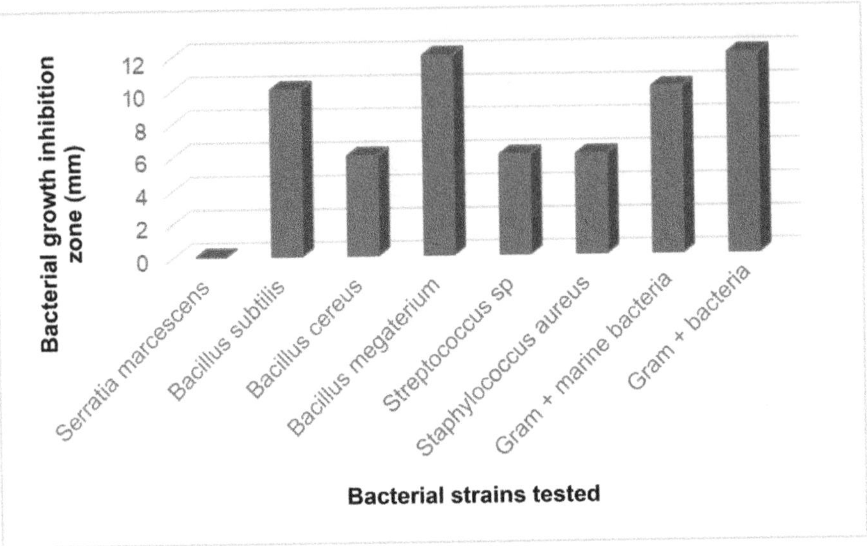

Figure 2.10. *Antibacterial activity of an ethanolic extract of* Enteromorpha intestinalis (Ulva intestinalis) *(according to Hellio et al. (2000))*

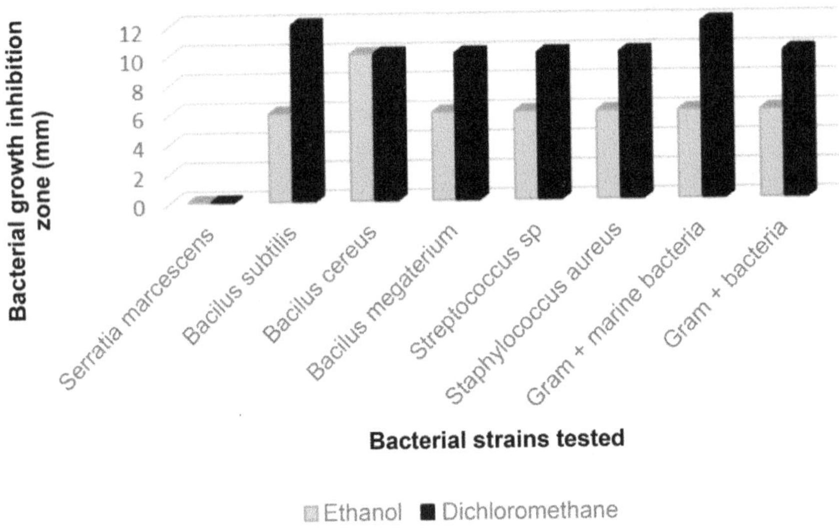

Figure 2.11. *Comparative antibacterial activity of ethanolic and dichloromethane fractions extracted from the red alga* Laurencia pinnatifida (Osmundea pinnatifida) *(according to Hellio et al. (2000))*

However, it is less effective at inhibiting the growth of Staphylococcus strains, particularly *Staphylococcus aureus*. Fractions obtained by extraction with ethanol or dichloromethane from certain green algae (*Chlamydomonas reinhardtii*) or brown algae (*Sargassum muticum*, *Ascophyllum nodosum*) proved ineffective in inhibiting bacterial growth. This suggests that the antibiotic activity of the extracts is associated with certain algal species, rather than all those that could potentially be used.

In some species, such as *Laurencia pinnatifida*, now known as *Osmundea pinnatifida*, both types of extract (ethanolic and dichloromethane) have antibacterial activity. However, the greatest activity is associated with the dichloromethane fraction (see Figure 2.11). In this species, the compounds that develop an antibiotic activity are probably hydrophobic in nature, with a more or less developed apolar character.

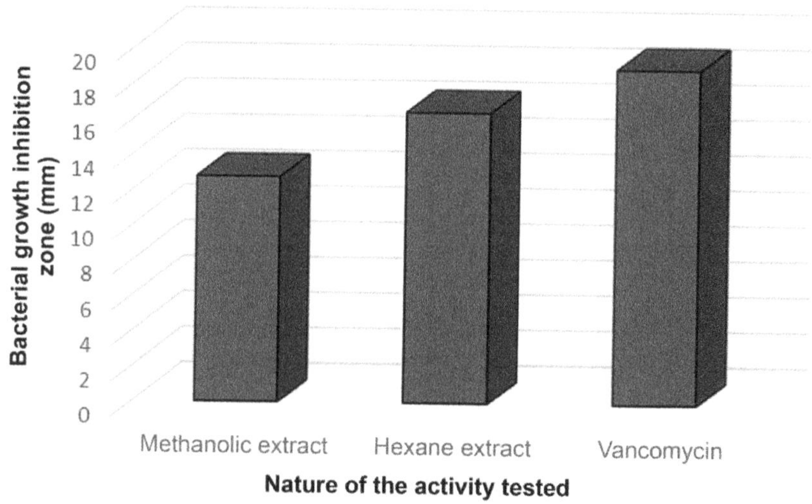

Figure 2.12. *Comparison of the antibacterial activity of different* Ulva intestinalis *and an antibiotic (Vancomycin) against a resistant strain of* Staphylococcus aureus *(MRSA 001 R) (according to Srikong et al. (2017))*

A more recent study shows that methanolic extracts of *U. intestinalis* are also endowed with antibacterial activity, notably against a methicillin-resistant strain of *S. aureus* (Srikong et al. 2017). This activity is associated with the methanolic fraction, as well as with the hexane fraction. This implies that this activity is distributed on polar compounds and on strongly hydrophobic substances. The inhibitory activity of the hexane fraction is close to the activity of vancomycin, an antibiotic commonly used against this resistant strain of *Staphylococcus aureus* (see Figure 2.12).

As previously mentioned, algal extracts appear to be more active on Gram+ bacteria than on Gram– bacteria. This is probably due to structural differences between the walls of the two bacterial types (see Figure 2.13). The wall of Gram– bacteria is more complex and probably less sensitive to the penetration of antimicrobial compounds.

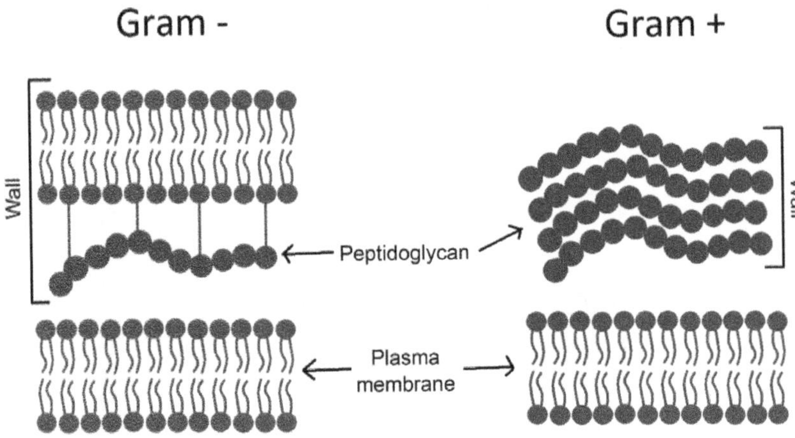

Figure 2.13. *Difference in the wall structure between Gram+ and Gram– bacteria (photo credit © Moon Rabbit 365 2017)*[1]

The bacterium *Bacillus subtilis*, often used in studies on the antibacterial activities of algal extracts, offers several advantages. It is a ubiquitous bacterium, and is considered a model species for the study of Gram+ bacteria. Although not pathogenic for humans, it can contaminate foodstuffs and sometimes cause food poisoning. In this context, the avoidance of antimicrobial substances in edible algae represents an asset for the development of this marine resource in human nutrition.

1 See: https://commons.wikimedia.org/w/index.php?curid=63943597.

A study carried out on over 20 species of algae showed that ethanolic or mixed extracts (chloroform/methanol) inhibited the growth of *B. subtilis*. This activity is particularly high with species of the genus *Caulerpa* (see Table 2.2) (Freile-Pelegrin and Morales 2004).

Species	Ethanolic extract	Chloroformic/methanolic extract
Caulerpa ashmeadii	10.3	11.3
Caulerpa cupressoides	11	-
Caulerpa prolifera	9.6	10
Caulerpa racemosa	7.3	7
Udotea occidentalis	12	14
Sargassum filipendula	-	7.7
Gracilaria caudata	12.3	10
Gracilaria cornea	8.3	8.7
Eucheuma isiforme	-	-
Digenea simplex	7.3	7.0

Table 2.2. *Antibiotic activity of different algal extracts (diameter of growth inhibition zone in mm) against the gram+ model bacterium* Bacillus subtilis *(according to Freile-Pelegrin and Morales (2004)).*

2.3. Antifungal activity

Candida albicans is a pathogenic fungus causing numerous infections of mucous membranes in humans. Certain methanolic algae extracts have been shown to have antifungal activity against this species, as well as against other strains (Arumugam et al. 2019). The extract most active against *C. albicans* is that obtained from the brown alga *Dictyota bartayresiana* (see Figure 2.14). The algal extract shows antifungal activity against other *Candida* strains, but with less efficacy (see Figure 2.15).

In the case of this alga, as with the species tested in this study, the extracts' mechanism of action appears to be fungistatic, not fungicidal (Armugan et al. 2019).

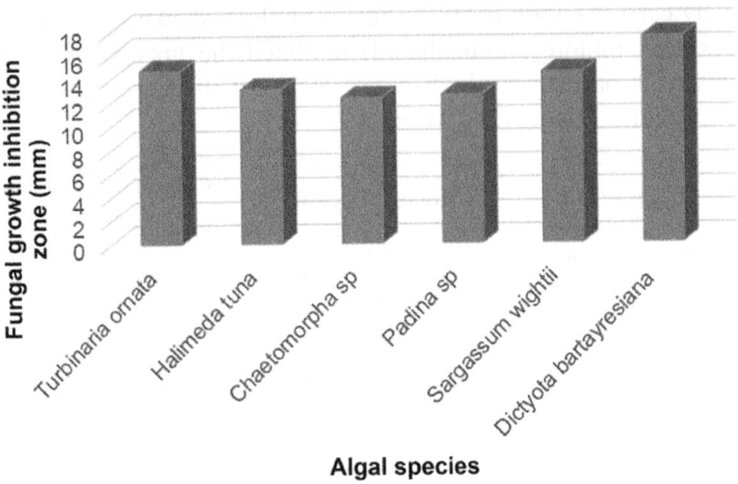

Figure 2.14. *Effect of methanolic extracts of different algal species on growth inhibition of the pathogenic fungus* Candida albicans *(according to Arumugam et al. (2019))*

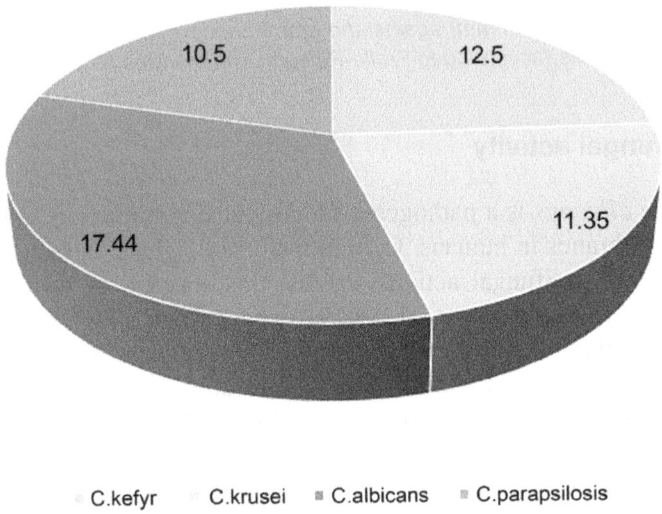

Figure 2.15. *Effect of a methanolic extract of the brown alga* Dictyota bartayresiana *on the inhibition (in mm) of the growth of different* Candida *strains (according to Arumugam et al. (2019)). For a color version of this figure, see www.iste.co.uk/fleurence/health.zip*

The antifungal action of algal extracts is not limited to *Candida* or *Saccharomyces* yeasts. Some algal species have also been described as having inhibitory activity against marine fungal strains. This is particularly true of the ethanolic extract of *Ascophyllum nodosum*, which is a powerful inhibitor of the growth of fungal strains from the marine ecosystem (Hellio et al. 2000). The presence of antifungal and antibacterial substances is also a factor to consider in order to better understand the in situ relationship between marine microbiota and algae.

2.4. Antiviral activities

The search for antiviral activities in algae extracts has been the subject of numerous studies and publications. The Google Scholar search engine, based on a query including the keywords *antiviral activity*, *seaweeds*, lists 23,700 references.

As in previous examples (see sections 2.2 and 2.3), the fractions tested are often obtained by extraction in the presence of methanol.

Some algal extracts have been shown to be active against human DNA viruses, such as polio (poliovirus type I) or herpes simplex (herpes simplex type I). They may also be active against RNA viruses, such as the Sindbis virus (Hudson et al. 1998). The latter virus is present in vertebrates and can affect humans, causing epidemic polyarthritis. The methanolic fraction of the green alga, *Codium fragile*, has been shown to be active against all three types of viruses. This alga is already used in traditional Korean medicine under the name "Cheong-gak" (see section 1.1.3). Of the many species tested, it is the only one to show activity against both DNA and RNA viruses. The mechanism of antiviral action may be direct, i.e. inactivation of the virus, or indirect, i.e. inhibition of its penetration or replication in cells. In most cases, the algal extracts tested showed virucidal, and therefore direct, activity. This is particularly true for the herpes simplex virus, where the minimum concentration of *C. fragile* extract required for inactivation is of the order of 4 µg/mL (see Figure 2.16). This low concentration required for virus inactivation is a good indicator of the extract's virucidal activity.

Extracts of the brown algae *C. bullosa* and *S. lomentaria* were found to have more moderate virucidal activity. However, at high concentration (250 µg/mL), they also provide cell protection against virus penetration (see Figure 2.16).

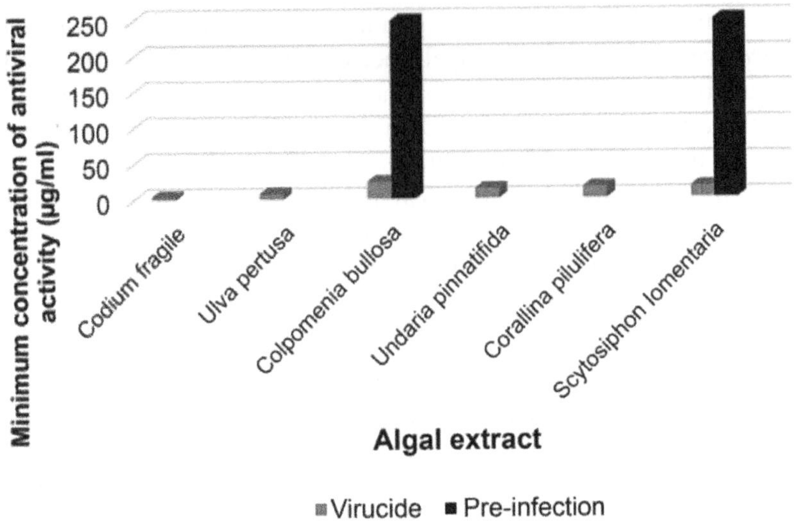

Figure 2.16. *Antiviral effect of different algal extracts on herpes simplex virus type I (according to Hudson et al. (1998))*

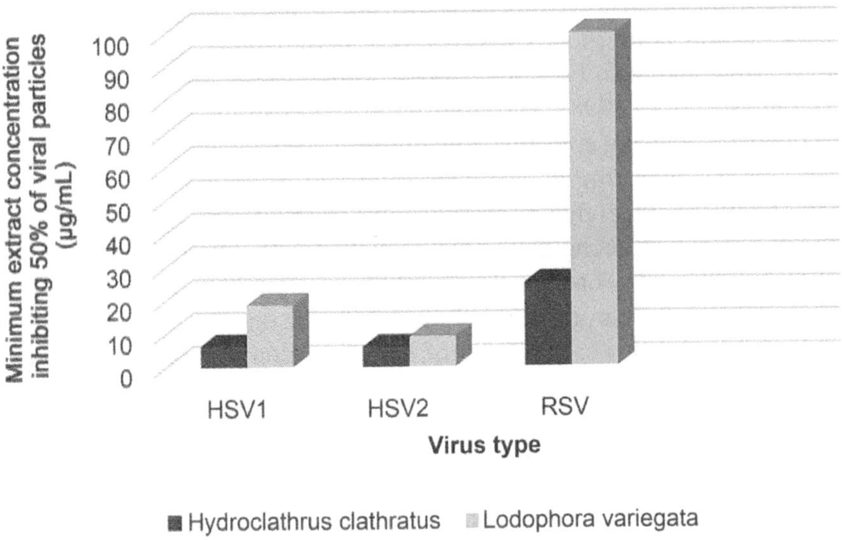

Figure 2.17. *Antiviral activity of an aqueous extract obtained under heat from two Korean brown algae* Hydroclathrus clathratus, Lobophora variegata *(HSV1: herpes simplex virus type I; HSV2: herpes simplex virus type II; RSV: respiratory syncytial virus) (according to Wang et al. (2008))*

The production of algal extracts with antiviral properties can sometimes be obtained using easy-to-implement, low-cost processes. This type of process is generally based on boiling water extraction. This extraction, applied to certain brown algae (*Hydroclathrus clathratus, Lobophora variegata*), enables the recovery of an active fraction, notably against herpes viruses (HSV1 and HSV2) and respiratory syncytial virus (RSV) (see Figure 2.17) (Wang et al. 2008).

Species	Reverse transcriptase inhibition
Dictyopteris jolyana (brown alga)	++
Dictyopteris polypodioides (brown alga)	++
Zonaria tournefortii (brown alga)	++++
Agardhiella ramosissima (red alga)	+
Gracilaria domingensis (red alga)	+
Osmundaria obtusiloba (red alga)	++++
Alsidium seaforthii (red alga)	++++
Botryocladia occidentalis (red alga)	+++
Codium isthmocladum (green alga)	+++

Table 2.3. *Inhibitory activity of aqueous extracts of HIV-1 reverse transcriptase from different algal species (according to Harb and Chow (2022))*

Some algal extracts have also been shown to be active against HIV-1 (see Figure 2.18) (Harb and Chow 2022). The algae concerned belong to the three major classes: Phaeophyceae (brown algae), Chlorophyceae (green algae) and Rhodophyceae (red algae) (see Table 2.3). The mechanism of action of these extracts is based on inhibition of HIV viral transcriptase. The active fractions are mainly obtained by methanolic or aqueous extraction. The concentrations required to achieve reverse transcriptase inhibition are

400 μg/mL for methanolic extracts and 200 μg/mL for aqueous extracts. This suggests that the aqueous fractions and the molecules solubilized in the latter are more active than those contained in the methanolic fraction.

This finding is confirmed by the determination of the minimum extract concentrations required for 50% inhibition of viral reverse transcriptase activity. The latter is generally very low with aqueous extract and much higher with methanolic extract (see Table 2.4).

Algal species	Aqueous extract IC50 (μg/mL)	Methanolic extract IC50 (μg/mL)
Zonaria tournefortii	< 5	315.32

Table 2.4. *Comparison of minimum concentrations required to inhibit 50% of viral reverse transcriptase activity between an aqueous and methanolic extract of the brown alga* Zonaria tournefortii *(according to Harb and Chow (2022))*

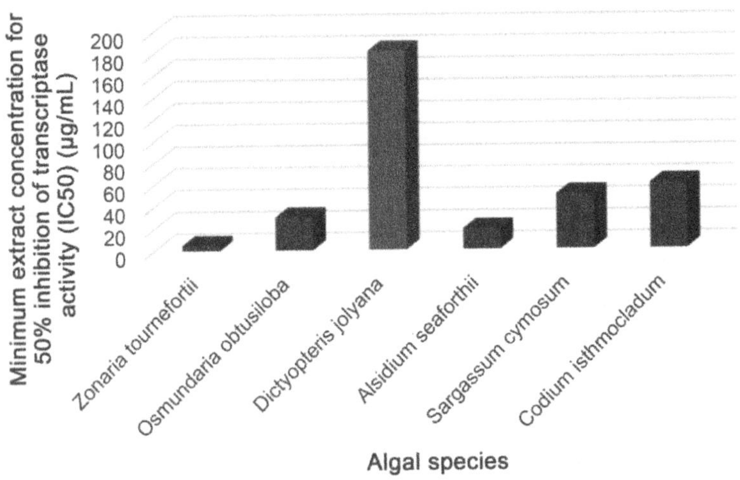

Figure 2.18. *Determination of the minimum concentration of algal extract to achieve 50% inhibition of HIV-1 reverse transcriptase activity (according to Harb and Chow (2022))*

The activity contained in a fraction obtained by hot aqueous extraction suggests that this activity is carried out by polar molecules, such as proteins, glycoproteins or anionic polysaccharides.

2.5. Anticoagulant and antithrombotic activity

Algal extracts, particularly from brown algae, are often cited for their anticoagulant activity. Fractions obtained by hot water extraction (30–100°C) of algae belonging to the genera *Undaria, Laminaria, Ecklonia* or *Sargassum* significantly increase the response time of thromboplastin (> 190 s), a coagulation factor (Kim et al. 2004). The most active fractions are reported for *Ecklonia and Laminaria* species (see Figure 2.19). In the alga *U. pinnatifida* (wakame), the observed activity differs depending on whether the extract is obtained from the whole alga (sporophyte) or from the stipe (see Figure 2.20). Differences in the recovered anticoagulant activity are also observed according to the type of fractionation following hot water extraction. The anticoagulant activity of *Ecklonia kurome* from hot water extraction and ethanol treatment results in two fractions, one of which is insoluble. Optimum activity is observed for the insoluble ethanolic fraction, suggesting that this activity is mainly carried out by complexes such as polysaccharides (see Figure 2.21).

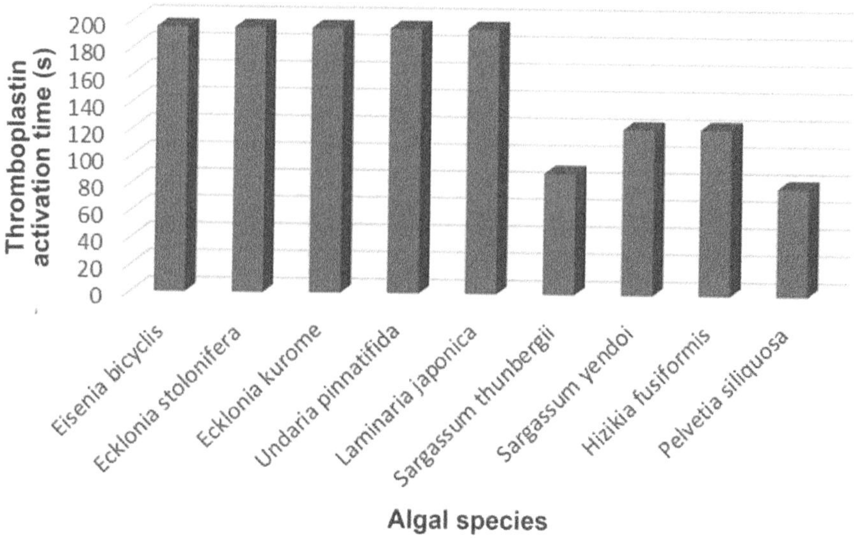

Figure 2.19. *Delayed activation of thromboplastin by brown algal fractions obtained by hot water extraction (according to Kim et al. (2004))*

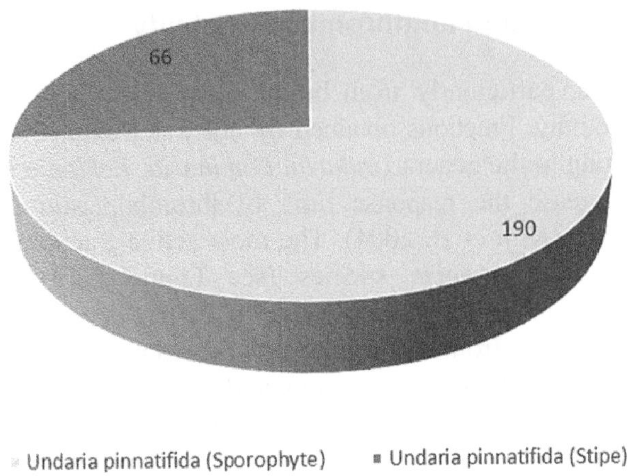

Figure 2.20. *Delaying activity (in seconds) of thrombinoplastin induction according to the parts of the brown alga* Undaria pinnatifida *(whole alga or stipe) (according to Kim et al. (2004))*

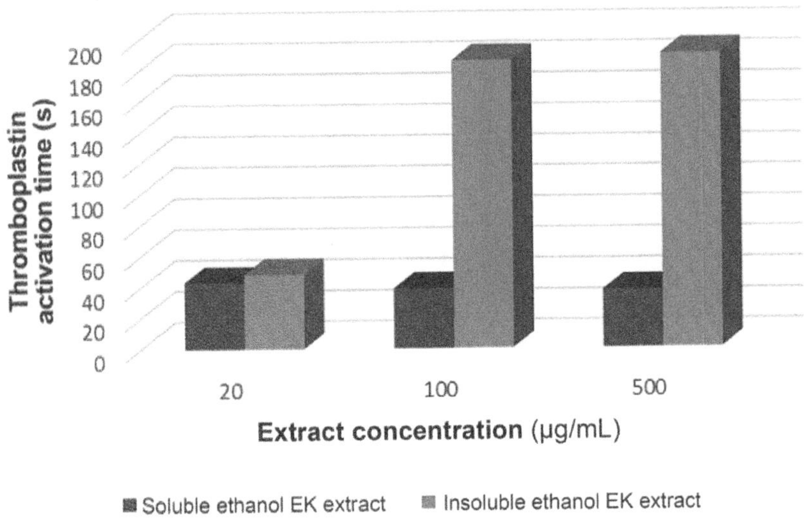

Figure 2.21. *Delaying activity of thromboplastin induction of soluble and insoluble ethanol fractions of the alga* Ecklonia kurome *(according to Kim et al. (2004))*

Thrombosis due to the formation of a blood clot is a serious vascular pathology, and the development of drugs with antithrombotic activity is a major focus of medical research. Clot removal relies on two mechanisms, thrombolysis or fibrinolysis.

Thrombolysis is the degradation by drugs of the clot obstructing blood vessels. Fibrinolysis, on the other hand, is a physiological process in which plasmin dissolves fibrin blood clots.

These two mechanisms are at the root of natural antithrombotic activity or induced antithrombotic activity.

Some algal extracts have been described as having antithrombotic activity. This is particularly true of the brown alga *Cladosiphon okamuranus*, frequently consumed in Japan under the name “mozuku” (see Figure 2.22) (Yasuzawa et al. 2019).

This activity was initially described in vitro on vascular endothelial cells. In this model, an algal extract showed an inducing effect on urinary plasminogen activating urokinase or u-PA. This enzyme induces the transformation of plasminogen into plasmin, a reaction usually inhibited by α 2-antiplasmin and α 2-macroglobulin. The plasmin thus synthesized ensures the proteolytic degradation of the fibrin constituting the clot (see Figure 2.23). Fibrinolysis is a complex but effective biochemical mechanism for eliminating thrombosis.

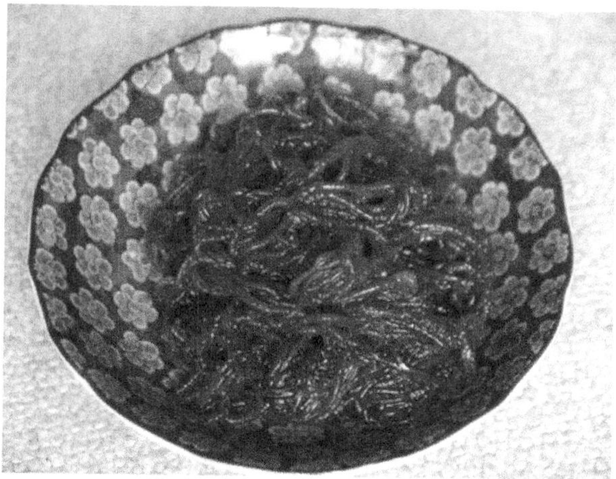

Figure 2.22. *Mozuku dish eaten in Japan* (Cladosiphon okamuranus) *(photo credit © https://commons.wikimedia.org/w/index.php?curid=17402345). For a color version of this figure, see www.iste.co.uk/fleurence/health.zip*

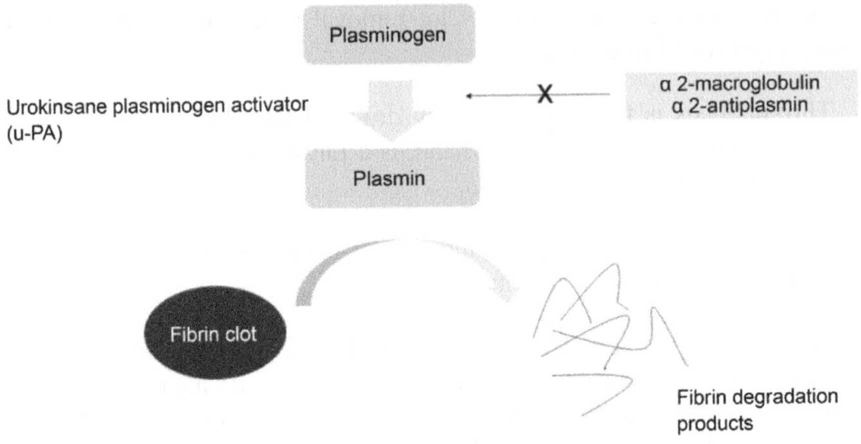

Figure 2.23. *Simplified mechanism of fibrinolysis*

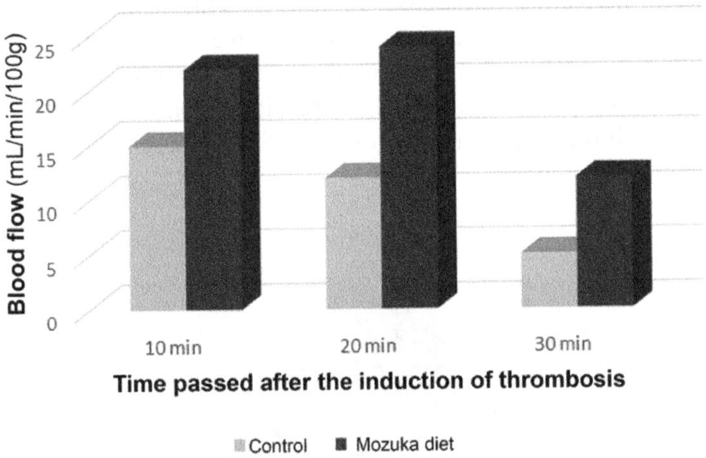

Figure 2.24. *Impact of mozuka extract consumption on the carotid blood flow in rats subjected to thrombosis induction via the injection of iron chloride ($FeCl_3$) (according to Yasuzawa et al. (2019))*

Mozuka extract administered at a concentration of 1% (w/v) increases the relative activity of u-PA by 50% (Yasuzawa et al. 2019) and thus represents a potential antithrombotic activity. The administration to rats of a diet incorporating mozuka extract also demonstrates an antithrombotic effect on the vascular system of treated animals via injection of a thrombosis-inducing ferric chloride solution ($FeCl_3$) (see Figure 2.24). After thrombosis induction, animals fed the mozuku diet notably showed a higher carotid blood flow than control rats affected by the same type of induced thrombosis.

Certain extracts of edible algae, and more particularly brown algae, are described in the scientific literature as having anticoagulant and/or antithrombotic activities. The fractions in question are obtained by extraction in polar solvents or hot water. This characteristic suggests that the molecules involved are carbohydrate-based. As will be shown later, many algal polysaccharides are associated with this type of activity, and in particular the fucans complex of brown algae, better known as fucoidans (see section 3.1.3).

2.6. Anti-inflammatory activities

Brown algae of the genus *Sargassum* are often used in traditional Chinese medicine for their antipyretic, analgesic and anti-inflammatory properties (see section 1.1).

The species *Sargassum fulvellum* and *Sargassum thunbergii* are described for the anti-inflammatory activity of their fractions obtained by extraction in an aqueous medium or in the presence of ethanol and dichloromethane (Kang et al. 2008).

Topical application of phorbol 12-myristate acetate (PMA) to the ear of mice elicited various inflammatory responses (edema, erythema, increased blood flow). The combined use of PMA and a dichloromethane extract of *S. fulvellum* (0.4 mg/ear) partially inhibited the onset of these inflammatory symptoms (see Figure 2.25). In this case, edema development was inhibited by 79%. By way of comparison, the classic anti-inflammatory drug indomethacin inhibited the same type of edema, induced under the same conditions on a mouse ear, by 83.7% (see Figure 2.26).

The same algal extract also had an inhibitory effect on erythema growth (24%) and on blood flow. In the latter case, blood flow was reduced by almost 40% compared to that observed in the presence of PMA and in the absence of sargassum extract.

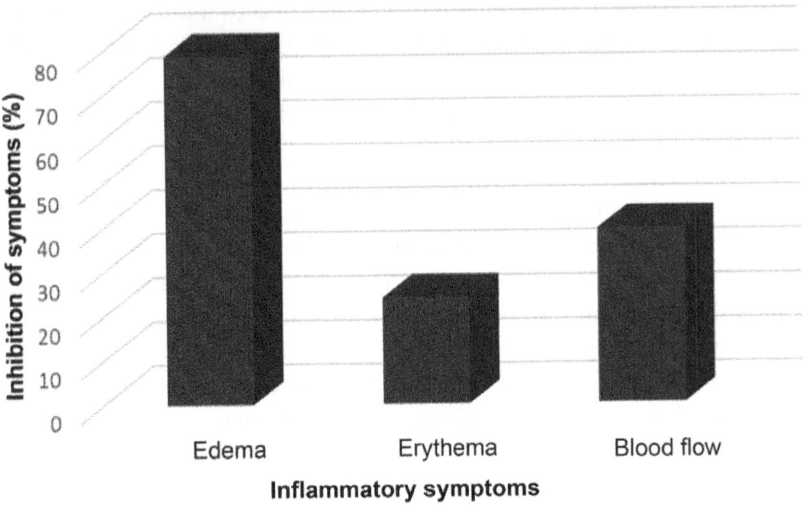

Figure 2.25. *Impact of applying an extract of* Sargassum fulvellum *(CH_2Cl_2) on dermal and blood inflammatory symptoms in a mouse treated with PMA(according to Kang et al. (2008))*

The same type of anti-inflammatory activity is reported with the dichloromethane extract of *S. thunbergii*. However, this activity is less intense than that reported with the *S. fulvellum* extract.

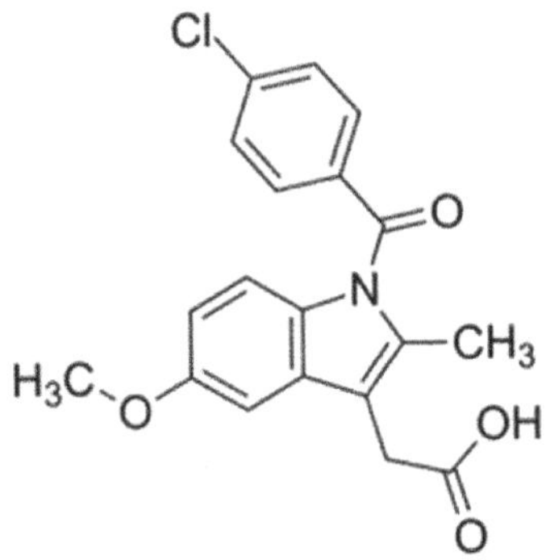

Figure 2.26. *Structure of indomethacin, the anti-inflammatory reference standard (photo credit © https://commons.wikimedia.org/wiki/File:2-(1-(4-chlorobenzoyl)-5-methoxy-2-methyl-1H-indol-3-yl)acetic_acid_200.svg#/media/File:2-(1-(4-chlorobenzoyl)-5-methoxy-2-methyl-1H-indol-3-yl)acetic_acid_200.svg))*

More generally, anti-inflammatory activity in the Sargassum algae studied appears to be more closely associated with the organic extract (CH_2Cl_2) than with the aqueous extract. This finding suggests that anti-inflammatory activity in brown algae of the genus *Sargassum* is linked to molecules with low water solubility. Phenols or their derivatives, widely represented in these algae, are potential carriers of this type of biological activity (see section 3.4).

Anti-inflammatory activities have also been reported for species other than sargassum. Methanolic extracts of *Ulva linza* (Chlorophyceae) or *Undaria pinnatifida* (Phaeophyceae) also show anti-inflammatory activity (Khan et al. 2008). At a dose of 40 mg/mL, the extract of these two algae inhibited 70–78% of PMA-induced edema in mouse models. The *U. pinnatfida* extract proved highly effective, with relatively low minimum inhibitory concentrations of 50% of symptoms, compared with *U. linza* (see Table 2.5).

Inflammatory symptoms	**IC_{50}** ***U. pinnatifida***	**IC_{50}** ***U. Linza***
Edema	10 mg/mL	20 mg/mL
Erythema	15 mg/mL	26 mg/mL
High blood fluid	18 mg/mL	31 mg/mL

Table 2.5. *Concentration of the algal extract required for 50% inhibition (IC_{50}) of inflammatory symptoms in mice treated with PMA (according to Khan et al. (2008))*

A dose effect was observed with the two algae extracts. Their effect on edema reduction increased with the dose administered (see Figure 2.27). At a maximum dose of 80 mg/mL, the *U. pinnatifida* extract almost totally inhibited induced edema. At this concentration, the *U. linza* extract strongly inhibited edema, but it was still present to the extent of 20%. This result confirms that the methanolic extract of *U. pinnatifida* is more active than that of *Ulva* in the treatment of the inflammatory symptoms tested. This finding applies with the same intensity and consequences to the treatment of erythema.

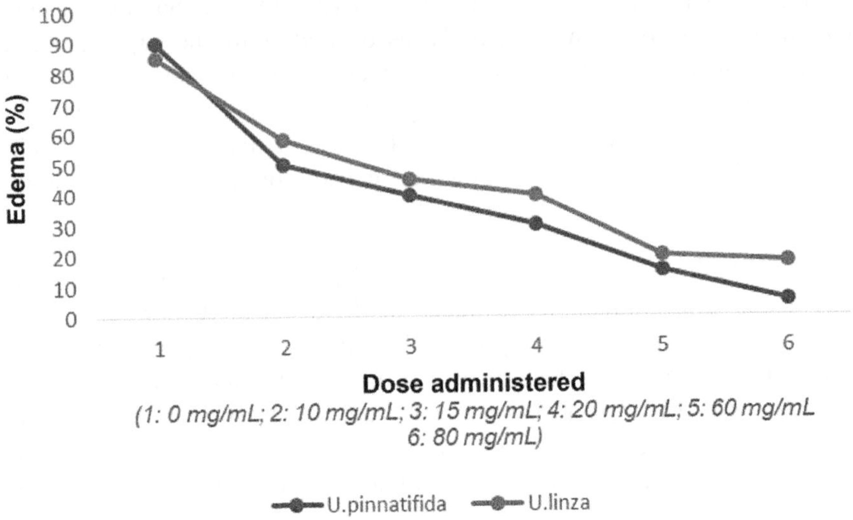

Figure 2.27. *Effect of applying increasing concentrations of methanolic extracts of* Undaria pinnatifida *and* Ulva linza *on edema in mice previously pre-treated with PMA (according to Khan et al. (2008)). For a color version of this figure, see www.iste.co.uk/fleurence/health.zip*

In *Undaria pinnatifida*, anti-inflammatory activity is also reported for ethanolic extracts. Application of an extract at a concentration of 0.4 mg per ear inhibited 95.3% of ear edema (see Figure 2.28) (Cho et al. 2007). The same type of activity has been described for the brown alga *Laminaria japonica*, a food known as "kombu" or "konbu" in Japan.

Fever is a physiological consequence of certain inflammatory processes, and many seaweed extracts are also described for their antipyretic activity. This property is highly valued in traditional Asian medicine, particularly Chinese (see section 1.1.1).

In the mouse model, the administration of *Undaria pinnatifida* extracts reduced body temperature. However, the most remarkable effect was seen with *L. japonica* extracts (see Figure 2.29).

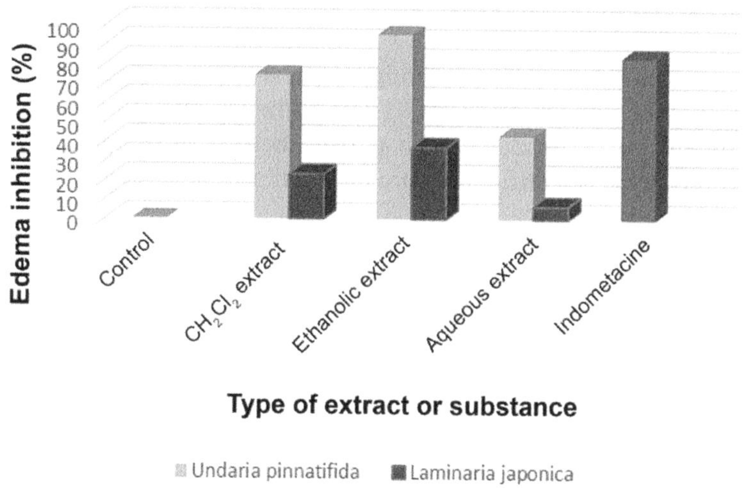

Figure 2.28. *Comparative efficacy of extracts of* Undaria pinnatifida *and* Laminaria japonica *in the inhibition of ear edema in the mouse model (according to Cho et al. (2007))*

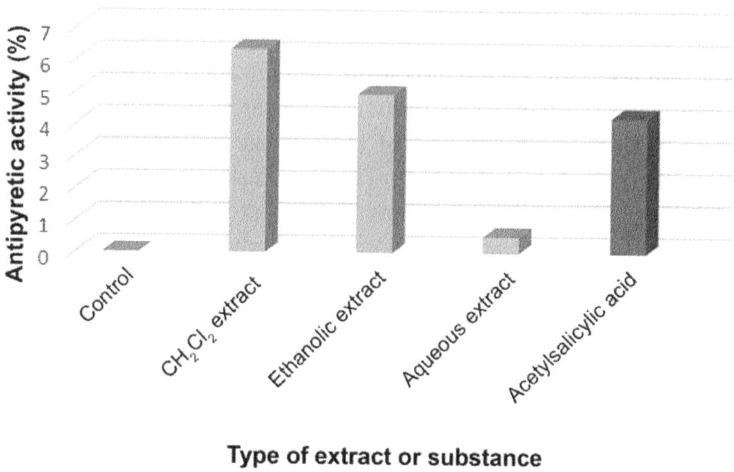

Figure 2.29. *Antipyretic effect of organic and aqueous extracts of the brown alga* Laminaria japonica *(according to Cho et al. (2007))*

2.7. Anti-allergenic activities

Allergy is an immune reaction triggered by a foreign substance known as an allergen. The allergenic reaction is part of the inflammatory mechanisms. Unlike many seafood products, such as fish, shellfish or seafood, there are a few citations in the scientific literature on the allergenic properties of algae.

On the PubMed or Web of Science scientific databases, cross-referencing the keywords Allergy–Human–Seaweed yields 39 and 27 published references respectively (James et al. 2023). A severe case of food poisoning following the consumption of ogonori (*Gracilaria verrucosa*) is reported in the scientific literature (Noguchi et al. 1994). However, this case could not be linked to a food allergy and is still, to this day, considered to be food poisoning due to an excess of prostaglandins.

This information suggests that algae either lack major allergens or possess compounds that limit the allergic response. The use of plant extracts or plant molecules as anti-allergenic agents is a potential therapeutic avenue already mentioned in the scientific literature (Bellik et al. 2012). Some of the molecules involved, such as polyphenols, luteolin and quercetin, are also present in certain species of algae.

In this context, the characterization of anti-allergic activities in algal extracts has been the subject of particular research (Fleurence and Ar Gall 2016). The inhibition mechanisms of the allergic reaction by algal extracts are partially known (Vo et al. 2012). They are best described in terms of the type I allergic reaction, also known as "immediate allergic hypersensitivity" (see Figure 2.30). This allergic response is characterized in particular by a significant production of IgE.

Algal fractions act on three pathways of the type I allergic reaction. They inhibit the synthesis of cytokines (interleukins 1 and 2) by Th2 cells (1). These cytokines are notably involved in the humoral immune response, i.e. in the activation of IgE-producing B lymphocytes. They are also thought to limit IgE production by B lymphocytes (2). Finally, algal extracts are thought to prevent mast cell degranulation (3). This process is responsible for the release of chemical mediators, such as histamine, the molecule responsible for many allergic symptoms (erythema, edema) (3) (see Figure 2.30).

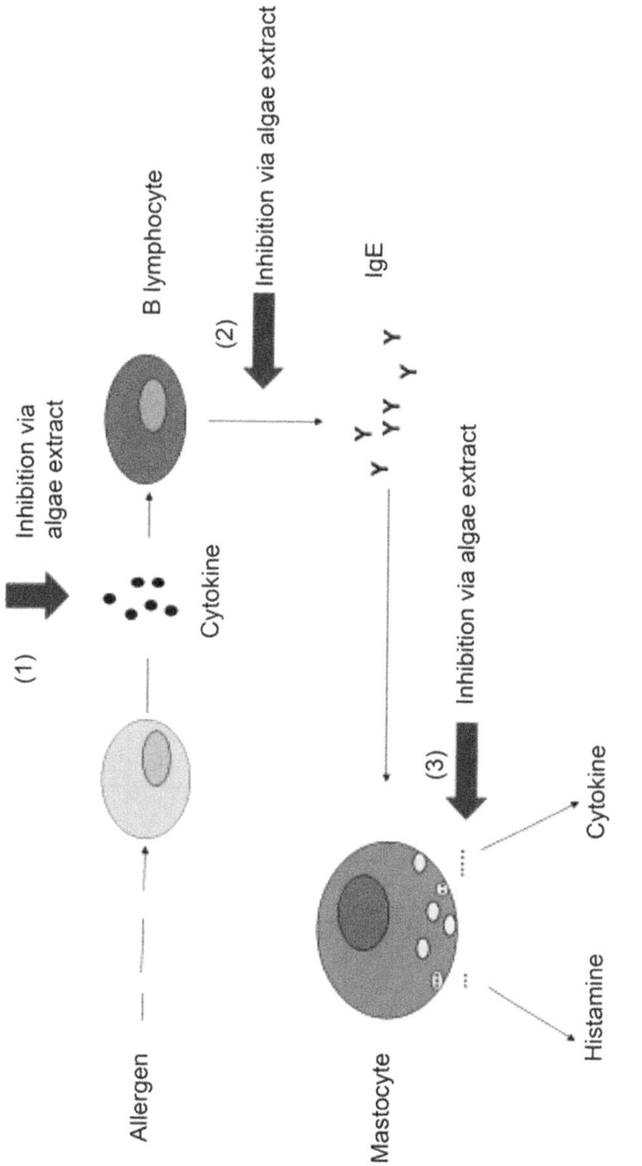

Figure 2.30. *Potential targets of the inhibitory activity of an algal extract on the general mechanism of a type I allergic reaction reaction (after Vo et al. (2012))*

Eisenia, *Ecklonia* and *Sargassum* algal extracts are described as having activity on the molecular mechanisms of food allergies. *Petalonia binghamiae* extract inhibits the degranulation mechanism behind mast cell cytokine release (see Figure 2.30). Fractions of the algae *Eisenia arborea* and *Sargassum thunbergii* reduce the level of histamine released, thus limiting the allergic inflammatory reaction. In the rat model, *Eisenia arborea* extract suppresses IgE production and very significantly lowers the level of histamine released (Vo et al. 2012).

The fractions tested are generally obtained by alcoholic (methanol or ethanol) or water extraction. Anti-allergic activities are mainly associated with brown algae (Phaeophyceae), as well as with a few species of red algae (Rhodophyceae) or green algae (Chlorophyceae) (see Table 2.6) (Fleurence and Ar Gall 2016).

Species	Country	Edible alga	Active extract
Ecklonia cava (brown alga)	Japan	Yes	Methanol
Eisenia arborea (brown alga)	Japan	Yes	Methanol
Eisenia bicyclis (brown alga)	Japan	Yes	Methanol
Sargassum thunbergii (brown alga)	China Japan	Yes	Ethanol Methanol
Sargassum cervicone (brown alga)	Pakistan	Yes	Ethanol
Saccharina japonica (brown alga)	China	Yes	Ethanol
Codium fragile (green alga)	Japan	Yes	Methanol
Ulva japonica (green alga)	Japan	Yes	Methanol
Porphyra dentata (red alga)	Japan	Yes	Methanol

Table 2.6. *Some examples of algal extracts belonging to the three main botanical groups (brown, green or red algae) whose extracts present anti-allergic activities (according to Fleurence and Ar Gall (2016))*

The molecules involved in this type of activity are often phenolic in nature and belong to the eckol family. Their structures will be detailed later (see section 3.4).

2.8. Antitumor activity

The presence of antitumor activity in an aqueous algae extract has been known for some 50 years (Nakazawa 1974).

Oral administration of dietary alga powder suspensions to mice affected by Ehrlich's carcinoma showed moderate to high efficacy against this dermal cancer (Noda et al. 1990) (see Figure 2.31). The powder was incorporated into the diet of mice at a rate of 1.6 g per kg murine body mass. Many species of brown algae in the Japanese consumer diet proved effective in the model studied. This was particularly true of the species *Laminaria japonica*, or kombu, frequently consumed in Japan. This is also the case of Hizikia fusiformis, or hijiki, whose tumor inhibition activity is close to 50% (see Figure 2.31). The highest rate of inhibition, however, was observed for the brown alga *Lessonia nigrescens* (60%), a species mainly used for the extraction of phycocolloids, such as alginates. In addition to brown algae, certain species of green and red algae also display in situ antitumor activity against Ehrlich's carcinoma. The Rhodophyceae *Porphyra yezoensis*, well known to consumers as "nori", shows tumor-inhibiting activity of approximately 53% (see Figure 2.31). The green alga *Ulva prolifera*, formerly *Enteromorpha prolifera*, also shows in situ antitumor activity in excess of 50%. This Chlorophyceae is one of a group of green algae, consumed singly or in mixtures, under the common name "aonori".

Some of the species studied for their inhibitory effect on Ehrlich carcinoma are also being studied for their antitumor activity on another type of cancer. This is Meth-A fibrosarcoma, a connective tissue cancer. In this new model, algal powder is suspended in saline and injected intraperitoneally into mice. The highest antitumor activity was obtained for the species *Sargassum patens* (55.5%) and lowest for *Monostroma nitidum* (12.3%) (see Figure 2.32).

As before, antitumor efficacy differs according to the species tested.

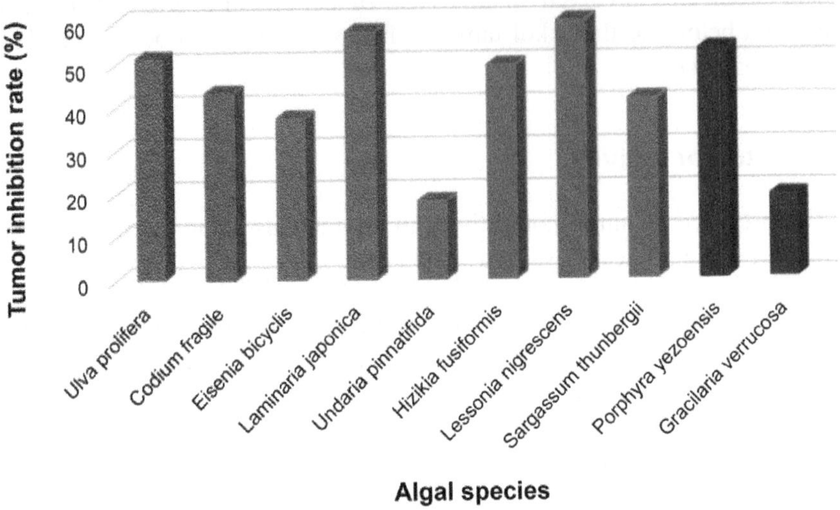

Figure 2.31. *Inhibitory effect of the oral administration of algal powder belonging to the three major algal botanical classes (Chlorophyceae, Phaeophyceae, Rhodophyceae) on the Ehrlich carcinoma tumor inhibition (1,600 mg powder/kg body mass) (according to Noda et al. (1990))*

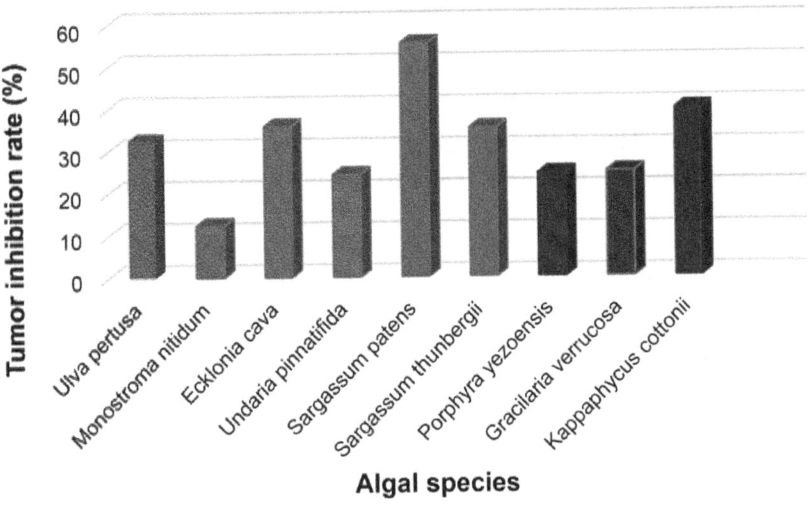

Figure 2.32. *Effect of injecting mice with algal solutions on the in situ inhibition of Meth-A fibrosarcoma (according to Noda et al. (1990))*

From one model to another and from one type of cancer to another (Ehrlich carcinoma or fibrosarcoma), the results obtained did not lead to the selection of a particular species. The mode of administration, i.e. oral or intraperitoneal injection, probably plays an important role in the efficacy of the treatments tested.

Apart from the search for antitumor activity using algal powders in suspension, other types of products have also been studied. These include algal fractions extracted with boiling water over a long period (4 h) (Yamamoto et al. 1982). The extract obtained was concentrated and dialyzed. The non-dialyzed fraction was then tested for its antitumor activity (see Figure 2.33). Its efficacy against L-1210 leukemia in mice was assessed in terms of life expectancy. Increased life expectancy was the main criterion for assessing the efficacy of algal extracts. Cancer cells and algal extracts were administered by intraperitoneal injection.

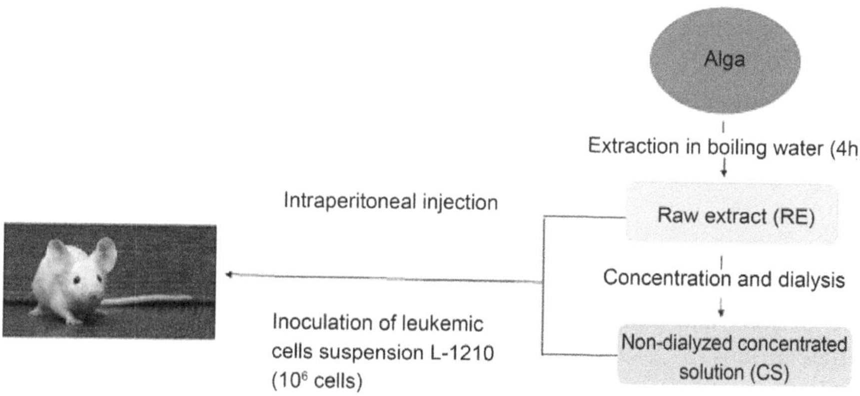

Figure 2.33. *Simplified procedure for testing algal extracts on the mouse model previously subjected to the injection of leukemic mice of the L-1210 strain (according to Yamamoto et al. (1982))*

Most algal extracts improved the life expectancy of mice affected by the inoculated tumor line (see Figure 2.34). Particularly active extracts were obtained from the green alga *Monostroma nitidum* and various brown algae of the genus *Ecklonia*, *Eisenia* and *Laminaria*. However, the extract of *L. angustata*, the longissima variety, proved to be active, while the extract of the same species, without determining the variety, proved to be inactive.

This particular point demonstrates, if proof were needed, the importance of taxonomic determination in biological screening.

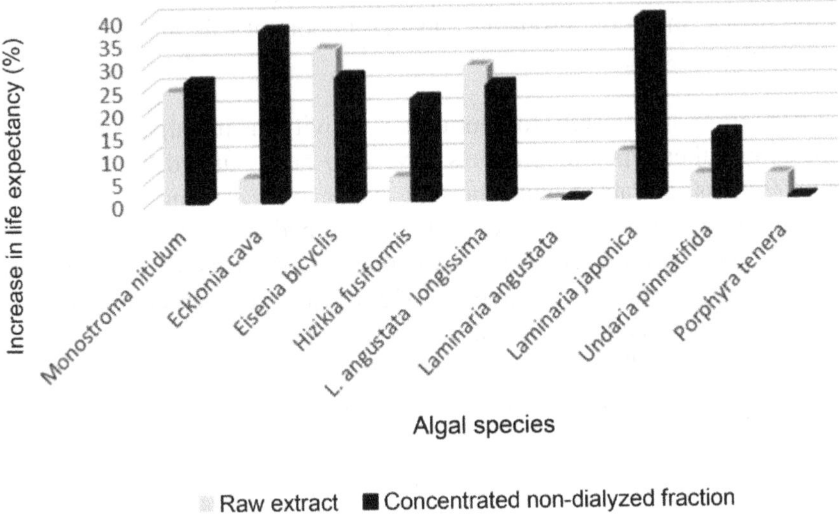

Figure 2.34. *Evaluation of the increase in life expectancy of leukemic mice (L-1210 strain) after inoculation with algal extracts (raw extract, concentrated non-dialyzed fraction) (according to Yamamoto et al. (1982))*

In many cases, the concentrated, non-dialyzable algal fraction proves more active than the starting crude extract (see Figure 2.34). This fraction, free of ions and small molecules, consists mainly of carbohydrates with sulfate groups (51.2–98.2%).

This suggests that the molecules with antitumor activity are of high molecular weight and carbohydrate nature. Numerous studies on the anti-cancer properties of algae highlight the algal polysaccharide fraction. The origin and nature of the polysaccharides involved may vary, as described later (see section 3.1).

2.9. Anti-cholesterol activity

In Western countries, the prevalence of cardiovascular disease as a cause of death remains high. The presence of cholesterol in the blood, or at least excess cholesterol, is a risk factor for this type of pathology. The brown algae frequently consumed in Asia appear in the literature as regulators of blood cholesterol, and their use in the diet is an interesting alternative or therapeutic supplement. A recently published journal takes stock of the cholesterol-lowering activities of algal extracts or algal powders incorporated into the diet (André et al. 2021). These extracts are described for their effect on "bad cholesterol" (LDL) and "good cholesterol" (HDL), as well as on the triglyceride content.

For example, the brown alga *Undaria pinnatifida*, or wakame, taken as a dietary supplement in powder form at 100 g/kg body weight, lowers blood triglyceride and LDL levels. On the other hand, this type of diet significantly raises blood levels of good cholesterol. However, the conditions under which this diet is implemented call for quantities of algae to be ingested that are ill-suited to Western consumption practices.

The alga *Ecklonia cava*, consumed in Japan and Korea under the name "kajime" or "norokajime", is also described for its hypo-lipidimic activity. Its administration in capsule form, two times a day, lowers LDL-type blood cholesterol levels. Independently of these formulations, alcoholic algal extracts also have cholesterol-lowering properties. This is the case of the brown alga *Ecklonia stolonifera*, or turuarame, traditionally consumed in Japan as a sea vegetable. Its ethanolic extract, administered at a dose of 100 mg/kg bodyweight, helps lower LDL cholesterol and triglycerides. Apart from brown alga, other species are also described in the literature for the same type of activity. This is the case, for example, with *Ulva lactuca*, consumed worldwide under various names (aonori, fisherman salad, sea lettuce). The effect of administering an algal ethanolic extract on the blood cholesterol of rats fed a high-fat diet (HFD) was the subject of a recent study (Widyaningsih et al. 2016).

This study showed that the use of ethanolic ulva extract (200 mg/kg body weight) significantly reduces blood cholesterol levels (-33%) in rats fed a HFD (see Figure 2.35). This concentration of algal extract gives a better result than Simvastatin, the cholesterol-lowering drug which is conventionally recommended.

Several studies describe the effect of algae consumption on blood cholesterol levels. This type of argument supports the integration of algae into our Western diet. It places these sea vegetables in the category of super foods, otherwise known as "health foods", as is the case in Japan.

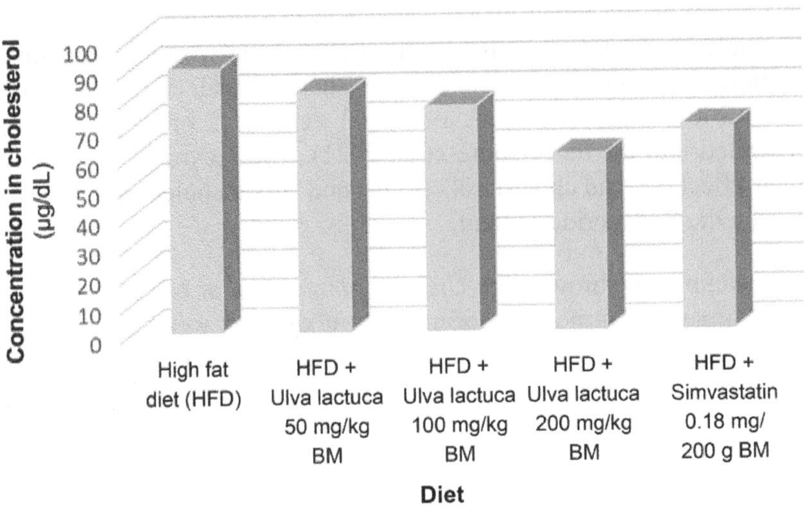

Figure 2.35. *Effect of the addition of an ethanolic extract of* Ulva lactuca *on blood cholesterol levels in rats fed a high-fat diet (HFD) (56 days). Comparison with the cholesterol-lowering drug Simvastatin (according to Widyaningsih et al. (2016))*

2.10. Other activities

Other activities of therapeutic interest associated with algal extracts include antiparasitic action against protozoa and worms.

With regard to protozoa, a screening study carried out on algae from the Normandy coast in France showed the presence of growth-inhibiting activities in the extracts. The protozoa targeted in this study were *Plasmodium falciparum*, the agent of malaria, *Trypanosoma cruzi*, the agent of sleeping sickness, and *Leishmania donovani*, the agent of leishmaniosis. The algal fractions tested were obtained by alcoholic extraction (ethanol) or with an apolar solvent (ethyl acetate) (Vonthron-Sénécheau et al. 2011). In almost 97% of cases, the antiparasitic activity was found in ethyl acetate extracts, suggesting that this type of activity is carried by apolar molecules.

The optimum activity of acetylated extracts was generally observed at a concentration of 9.7 µg/mL. At this concentration, the best parasite growth inhibition activity was observed with the extract of the brown alga *Bifurcaria bifurcata*. Inhibition was 100% for *Plasmodium* and *Leishmania*, and 78% for the trypanosome. Generally speaking, the antiprotozoal activity appeared to be associated more with brown algae, a little with red algae and very little with the green algae tested (see Figure 2.36).

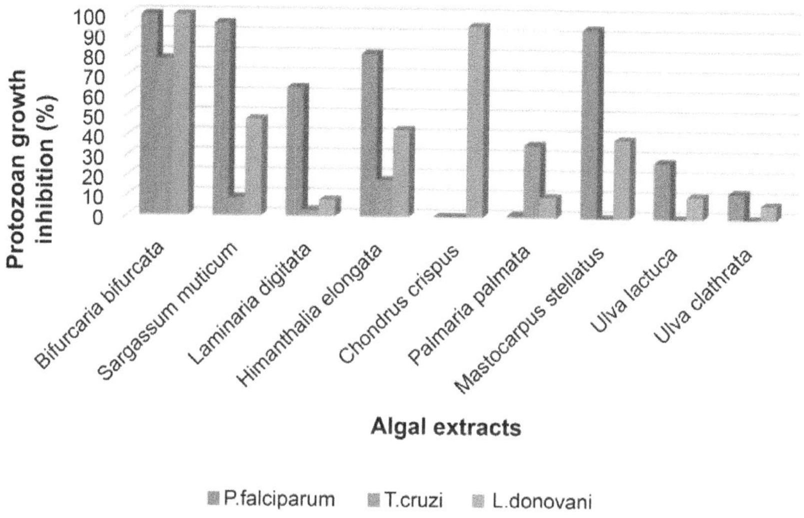

Figure 2.36. *Effect of algae extracts (ethyl acetate) on the growth inhibition of* Plasmodium falciparum, Trypanosoma cruzi *and* Leshmania donovani *(according to Vonthron-Sénécheau et al. (2011)). For a color version of this figure, see www.iste.co.uk/fleurence/health.zip*

Plasmodium falciparum is the protozoan most responsive to the action of algal extracts, with the exception of *Palmaria palmata* (see Figure 2.36).

The brown alga *Bifurcaria bifurcuta* has also been described in the literature for its antihelminthic activity against the nematode *Heligmosomoides polygyrus bakeri* (Miclon et al. 2023). This pathogenic worm causes intestinal infections in mammals.

The activity of aqueous extracts of this brown alga and its impact on the development of the worm are particularly well described. This impact is assessed through three phases in the life of the parasite: egg hatching, larval mortality after hatching and subsequent larval development. The nematicidal effect of *B. bifurcata*'s aqueous extract is particularly remarkable, since its application results in 67% larval mortality after the hatching process (see Figure 2.37).

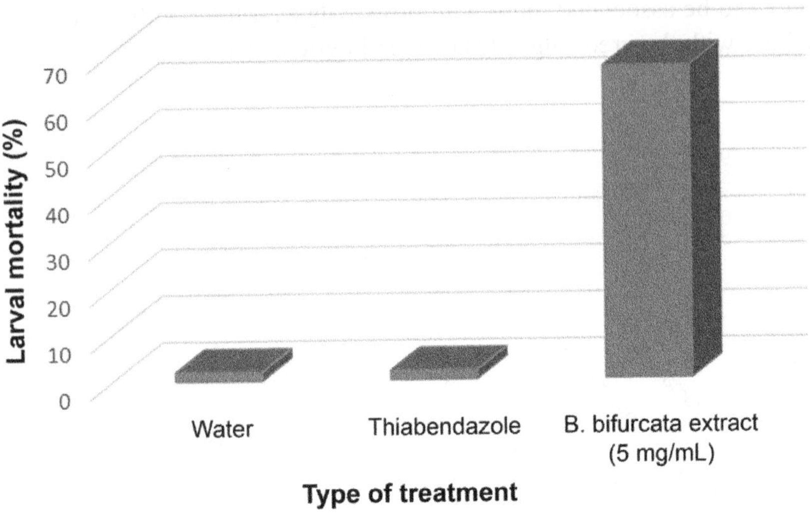

Figure 2.37. *Nematicidal effect of the aqueous extract of* Bifurcaria bifurcata *(5 mg/mL) on* Heligmosomoides polygyrus bakeri *(according to Miclon et al. (2023))*

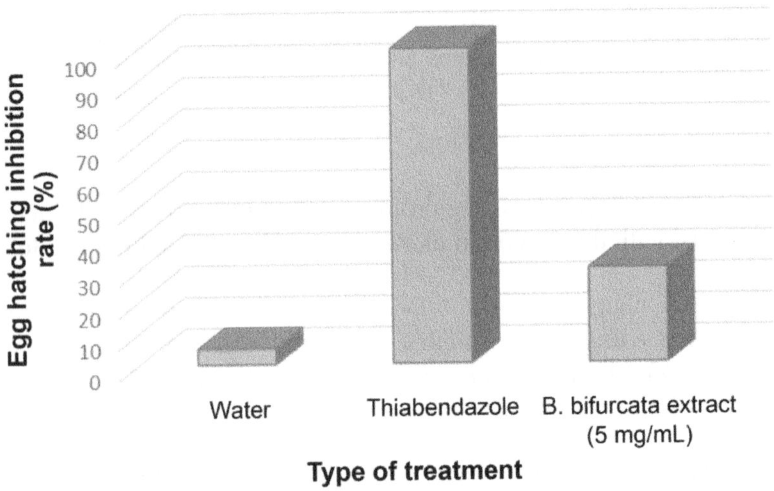

Figure 2.38. *Inhibitory effect of the aqueous extract of* Bifurcaria bifurcata *(5 mg/mL) on the hatching of* Heligmosomoides polygyrus bakeri *eggs (according to Miclon et al. (2023))*

Thiabendazole, a frequently recommended deworming treatment, is clearly less effective than the algal extract as a nematicide. This finding must be qualified, however, by the fact that thiabendazole is significantly more effective than the algal extract in inhibiting parasite egg hatching (see Figure 2.38). In addition to its effects on egg hatching and larval mortality, the *B. bifurcata* extract is particularly effective on parasite development. In fact, it inhibits 100% of the animal's development, on par with ivermectin, the pharmaceutical molecule used to treat this parasite (see Figure 2.39). On the other hand, the red algae extracts tested, *Grateloupia turuturu* and *Osmundea pinnatifida*, were not very effective in inhibiting the nematode.

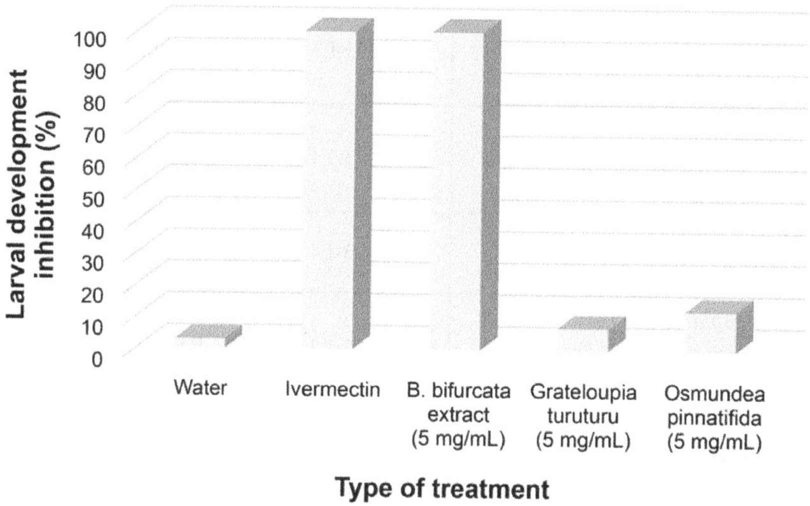

Figure 2.39. *Inhibitory effect of the aqueous extract (5 mg/mL) of* Bifurcaria bifurcata, Grateloupia turututuru, Osmundea pinnatifida *and ivermectin on the larval development of* Heligmosomoides polygyrus bakeri *(according to Miclon et al. (2023))*

In addition to the pathologies inherent in parasitic infection (protozoa, helminths), algal extracts have also been described for their effect on obesity (Gómez-Zorita et al. 2020). In particular, they are said to influence adipogenesis, the cellular process responsible for the transformation of pre-adipocytes into adipocytes and the storage of lipids in the latter. Numerous algal extracts from different classes (Phaeophyceae, Rhodophyceae, Chlorophyceae) have been shown to be active in this cell differentiation process (see Table 2.7).

Algal extract	**Employment concentration (µg/mL)**	**Cell line**	**Effect**
Sargassum thunbergii (brown alga)	10/50/100	373-L1	Reduced adipogenesis
Caulerpa okamurae (green alga)	25/50/100	373-L1	Reduced adipogenesis
Undaria pinnatifida (brown alga)	1/10	373-L1	Reduced adipogenesis
Gelidium amansii (brown alga)	50/100	373-L1	Reduced adipogenesis
Grateloupia elliptica (red alga)	25/50/100/200	373-L1	Reduced adipogenesis
Plocamium telfairiae (red alga)	25/50/100	373-L1	Reduced adipogenesis

Table 2.7. *Effect of algal extracts on adipogenesis in cells of the 371-L1 lineage (according to Gomez-Zorita et al. (2020))*

The anti-adipogenic molecular mechanisms of algal extracts are partly known. The algal extract of *S. thunbergii* strongly reduces the levels of mRNA and proteins involved in the activation of the peroxisome, a cellular organelle involved in lipid metabolism.

Numerous studies have highlighted the presence of various therapeutic activities associated with algal extracts. Many of these activities are associated with brown algae (*Sargassum* sp., *Laminaria* sp., *Undaria* sp., *Hizikia* sp.). This explains the empirical use of these species in Chinese, Japanese and Korean medicines (see sections 1.1–1.3).

The biological activities of these extracts are carried out by macromolecules or molecules that need to be identified, in order to complete the promotion of algae as a source of medicines from the sea.

3

Molecules

The above-mentioned therapeutic activities are characterized using algal extracts. This type of finding reinforces the importance of using algae in our diet and associating their use with a health prevention strategy.

These pharmacological activities, of undeniable interest in human health, are due to the presence of molecules, some of which are characteristic of algae. These include sulfated polysaccharides, certain glycolipids (GLs), polyphenols and protein pigments such as phycobiliproteins.

3.1. Polysaccharides

In algae, polysaccharides can play a structural or reserve role. In particular, they make up the cell wall and ensure its structure. One of the original features of these polysaccharides is their charged nature, generally carried by carboxylic, alcoholic or sulfated groups. This ionic character gives many algal polysaccharides a strong affinity for water, and therefore a high retention capacity for this liquid. This physico-chemical property enables algae thallus to retain water and thus withstand the daily periods of emersion that the foreshore undergoes.

3.1.1. *Alginates*

Brown algae are characterized by their high carbohydrate content. The species *Laminaria digitata* (Breton kombu) has a carbohydrate content ranging from 42 to 62% of the seaweed's dry matter (Darcy-Vrillon 1993).

These carbohydrates are mainly polysaccharides, particularly alginates, whose content varies between 20 and 26% of the plant's dry weight. Alginates are present in many species of brown algae (*Ascophyllum* sp., *Laminaria* sp., *Macrocystis* sp., *Fucus* sp., *Sargassum* sp., *Ecklonia* sp.). Species of the genera *Laminaria* (*Saccharina*), *Macrocystis* or *Ascophyllum* are the main sources used for the extraction of this polysaccharide. The species *Macrocystis pyrifera*, mainly exploited on the Californian and Chilean coasts, has a maximum alginate content of 21%, slightly below the maximum levels observed for kelp.

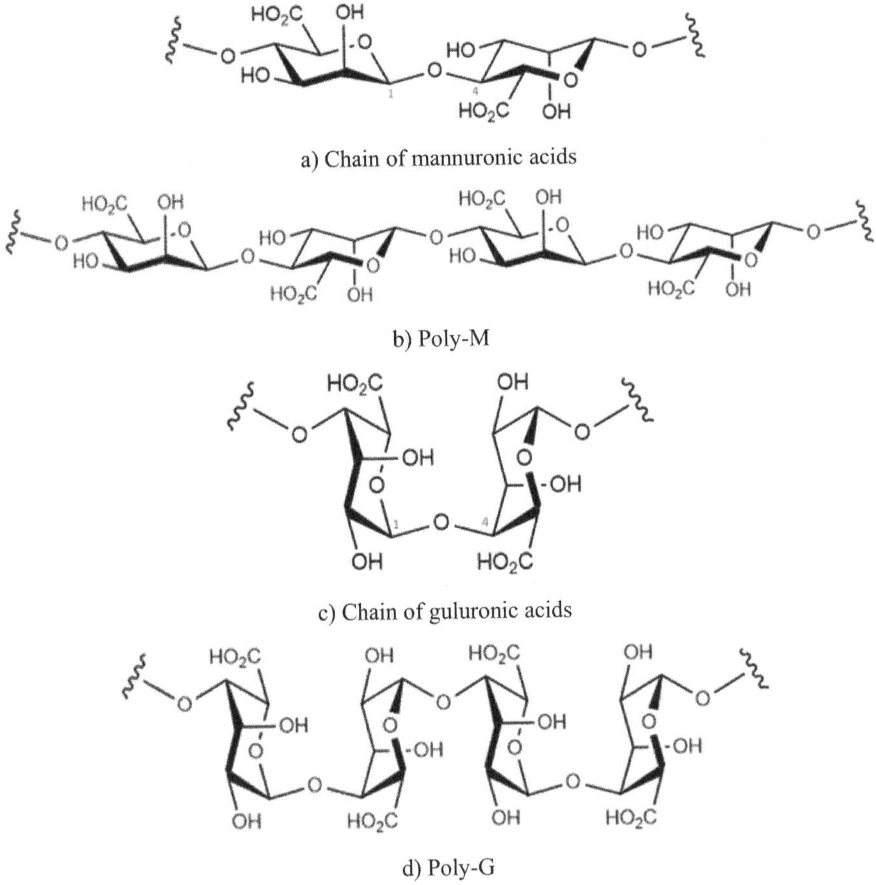

Figure 3.1. *Structure of the basic constituents of an alginic acid molecule (mannuronic acid-guluronic acid and poly-M, poly-G blocks) (according to Fleurence (2018))*

Alginate is the ionized form of alginic acid. Alginic acid is a polymer made up of mannuronic and guluronic acids (see Figure 3.1). This type of chain is not necessarily homogeneous, and there is often a succession of mannuronic (poly-M) and guluronic (poly-G) acid blocks. The frequency with which some of these blocks are present is responsible for the greater or lesser viscosity of alginates and hence their physicochemical properties.

Alginates are described as having numerous activities of therapeutic interest. These include antitumor, anti-inflammatory, antigastric reflux and antihemorrhagic (hemostatic) properties.

Alginates from the brown alga *Sargassum vulgare* have been described for their in vivo efficacy in inhibiting the growth of solid tumors belonging to Sarcoma 180, a connective tissue cancer. Algae of the genus *Sargassum* are well known to traditional Asian pharmacopoeia, which frequently recommends their use in the treatment of a wide variety of pathologies (see Chapter 1). In *Sargassum vulgare*, two types of alginate have been identified, distinguished by their molecular weight and, above all, their viscosity. These are low-viscosity alginates (LVAs) and high-viscosity alginates (HVAs) (de Sousa et al. 2007). LVAs are characterized by a molecular weight of 19.3 g/mol and a high proportion of mannuronic blocks, to the detriment of guluronic blocks. They are also characterized by an M/G ratio of 1.38, significantly higher than that observed for HVA-type alginates (1.04). Oral or intraperitoneal administration of these alginates to mice affected by Sarcoma 180 produced convincing results in terms of tumor growth inhibition (see Figure 3.2).

Alginates at a dose of 100 mg/d are particularly effective in inhibiting tumor growth. At this concentration, LVAs show an inhibitory effect of almost 75% and HVAs approximately 89% (see Figure 3.2). At these doses, *S. vulgare* alginates show comparable or even slightly greater activity than fluorouracil, the reference antitumor molecule for the treatment of sarcoma. The alginates' mode of administration also appears to play a role in antitumor efficacy. In particular, the oral route delivers the best results. This mode of administration generates tumor inhibitory activity some two times greater than that observed after intraperitoneal injection of alginates (de Sousa et al. 2007). According to the aforementioned authors, the anticancer activity of alginates could result from their ability to activate the immune system.

The antitumor activity of alginates from species other than sargassum is also well described. For example, alginates derived from *Laminaria brasiliensis* induce atypical mitoses on Hela cancer cells in vitro at low concentrations (µg/mL). As in the previous example, the ratio of mannuronic blocks to guluronic blocks seems to play an important role in the antimitotic, and therefore antitumoral, activity of the alginates studied (Khotimchenko 2010). These alginates are characterized by an M/G ratio of 1:2 and are active at a concentration of 160 µg/mL. This example suggests that alginates can act directly on cancer cell multiplication in vitro.

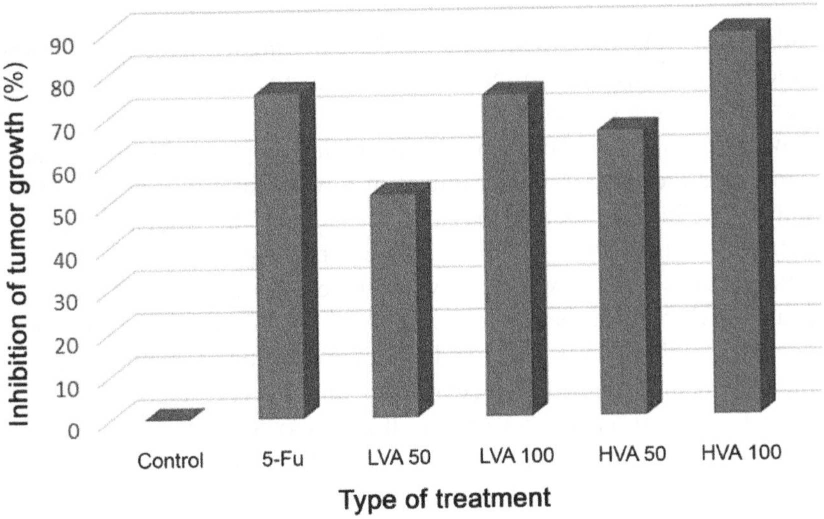

Figure 3.2. *Effect of oral administration of alginates of the brown alga* Sargassum vulgare *on in vivo inhibition of Sarcoma 180 tumor growth inoculated into a murine model*

COMMENT ON FIGURE 3.2 – *5-Fu: 5 fluorouracil; LVA 50: low viscosity alginates 50 mg/m²/d; LVA 100: low viscosity alginates 100 mg/m²/d; HVA 50: high viscosity alginates 50 mg/m²/d; HVA 100: high viscosity alginates 100 mg/m²/d (according to de Sousa et al. (2007)).*

Ehrlich's carcinoma is a study model often used to investigate the antitumor activity of algal extracts (see section 2.8). Anticancer activity in this type of dermal tumor also varies according to the type of alginate and the dose administered. Poly-M fractions appear more active than poly-G fractions at doses of 100 and 200 mg. A dose effect was also observed at an optimum concentration of 200 mg, irrespective of the type of alginate supplied orally to the mice (see Figure 3.3) (Noda et al. 1990).

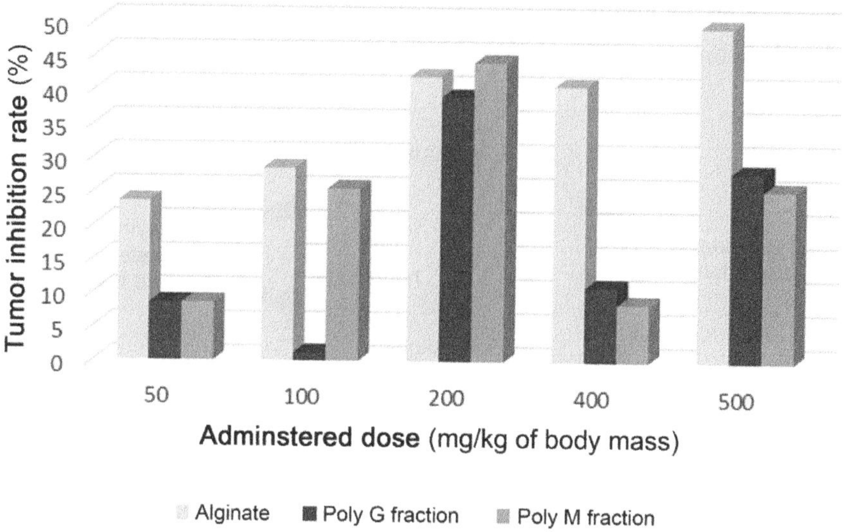

Figure 3.3. *Comparison of the antitumor activities of sodium alginate and derived poly-G and poly-U fractions against Ehrlich's carcinoma in the murine model (according to Noda et al. (1990))*

In addition to alginate, a natural polysaccharide found in many brown algae, research is also focusing on oligosaccharides alginates derived from the degradation of the initial polymer (Xing et al. 2020). These oligosaccharides, known as AOS for "alginate oligosaccharides", can be obtained by three different pathways (see Figure 3.4). One of these pathways is the enzymatic hydrolysis of the polysaccharide via a specific enzyme or alginate lyase present in many microorganisms, as well as in algae and certain marine invertebrates (Wong et al. 2000; Xing et al. 2020). Among these organisms are some brown algae, such as *Laminaria digitata* or *Undaria pinnatifida*, and some marine gastropods, such as abalone (see Table 3.1). Biochemically, alginate lyases fall into two distinct categories, according to their cleavage specificity towards sequences rich in mannuronic acids (poly-M) or guluronic acids (poly-G). Alginate lyases EC 4.2.2.3 are specific to poly-M sequences, while alginate lyases EC 4.2.2.11 are characteristically involved in the cleavage of poly-G sequences.

In addition to their catalytic class, these enzymes can also exist as isoenzymes. This is the case in the marine gastropod *Turbo cornutus*, or sea snail, which has two isoforms of alginate lyase called SP1 and SP2 (see Table 3.1). These two forms have the same cleavage specificity, but differ in their pH optimum zones.

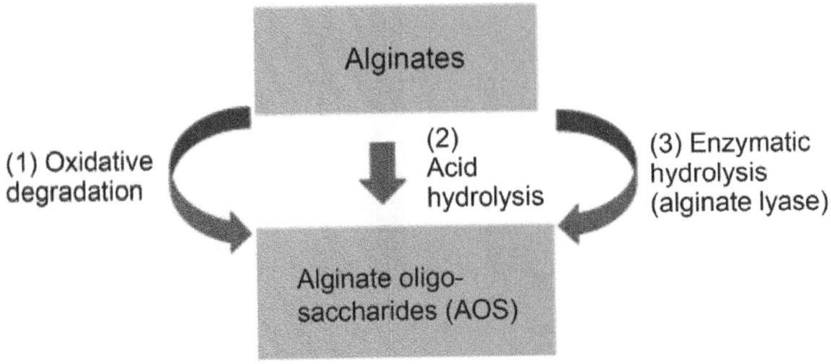

Figure 3.4. *Pathways for obtaining oligosaccharides alginates*

Organisms	Cleavage specificity	Optimum pH	Molecular mass (kDa)
Undaria pinnatifida (brown alga)	Poly-M	7.7	74
Laminaria digitata (brown alga)	Poly-M/ poly-G	-	-
Pelvetia canaliculata (brown alga)	Poly-M/ poly-G	7.5	-
Haliotis tuberculata (gastropod mollusk)	Poly-M	8.0	34
Aplysia depilans (gastropod mollusk)	Poly-M	9.6	-
Turbo cornutus (gastropod mollusk)	*SP1: M *SP2: M	*SP1: 7.5–9.2 *SP2: 7.7–9.4	32
***Vibrio* sp.** (marine bacteria)	Poly-M/ poly-G	8.5	40–42
***Pseudomonas* sp.** (bacteria)	Poly-M/ poly-G	7.5	32

*SP1: alginate lyase SP1; *SP2: alginate lyase SP2.

Table 3.1. *Examples of organisms with alginate lyases (according to Wong et al. (2000))*

AOS produced enzymatically, or by any other pathway, are often described for their antitumor properties. Their mechanisms of action are many and varied. AOS appears to be involved in inhibiting the proliferation and migration of cancer cells. Their oligosaccharidic nature is also thought to play a role in stimulating the body's immune and anti-inflammatory defenses (Xing et al. 2020). The role of AOS in limiting the invasiveness of tumor cells is sometimes put forward. This is particularly the case in prostate cancer, where AOS are described as activators of the Hippo signaling pathway. This metabolic pathway, which regulates cell proliferation and apoptosis, is part of the defense mechanisms against cancer (Han et al. 2019). This recent work, already cited more than 40 times, was the subject of a retraction procedure by its authors at the end of 2021. An update of the results now appears to be underway.

In a more consensual way, it is accepted that OAS play a role as immuno-modulators. For example, mannuronate and guluronate oligomers (20–24 units) at a concentration of 500 μg/mL increase the defense response against cells of the leukemia tumor line U937 (Xing et al. 2020). This response results in activation of the synthesis of toxic cytokines produced by blood mononuclear cells.

AOS, produced enzymatically using alginate lyase and subsequently transformed by chemical substitution, may also have antitumor activity.

Two AOS (A, B) chemically modified by the addition of sulfate groups have been shown to inhibit the Sarcoma 180 tumor implanted in a murine model (Hu et al. 2004).

Oligosaccharide A, with an apparent molecular mass of 3,798 Da and a degree of sulfation of 1.3, showed a tumor-inhibiting activity of 70% at a dose of 100 mg of product administered per kg of body mass (see Figure 3.5). At a dose of 50 mg/kg, the inhibitory effect of the oligosaccharide was only 66%, but this is still a significant result.

Oligosaccharide B, with an apparent molecular mass of 3,417 Da and a lower degree of sulfation (1.1), gave a lower net tumor inhibition rate than that reported with oligosaccharide A (see Figure 3.5). The presence of sulfate groups and the degree of sulfation of the oligosaccharide appear to be a determining element for the effectiveness of the product tested (see Figure 3.6). Indeed, at a dose of 100 mg/kg, AOS C and D, which have not undergone sulfation, are significantly less effective in inhibiting tumors than AOS A and B, which have undergone sulfation (see Figure 3.6).

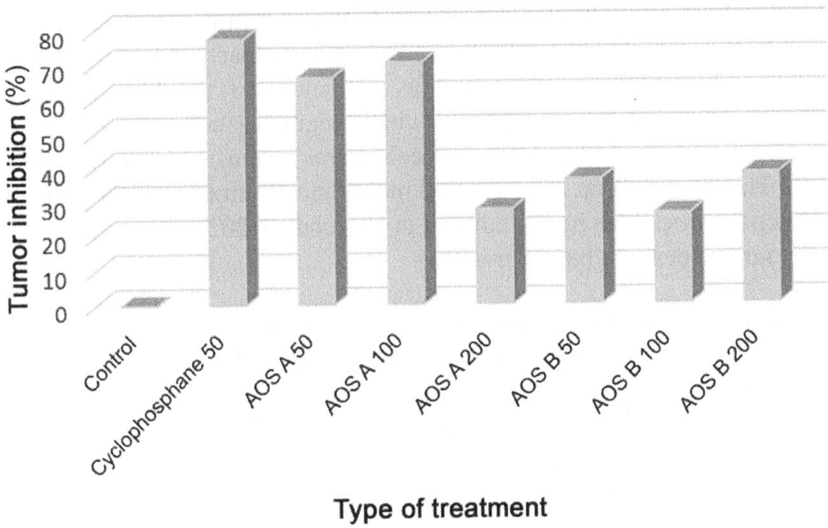

Figure 3.5. *Effect of administration of sulfated alginate oligosaccharides on the inhibition of the Sarcoma 180 tumor inoculated into a murine model (according to Hu et al. (2004))*

COMMENT ON FIGURE 3.5 – *AOS A 50: alginate oligosaccharide A 50 mg/kg; AOS A 100: alginate oligosaccharide A 100 mg/kg; AOS B 50: alginate oligosaccharide B 50 mg/kg; AOS B 100: alginate oligosaccharide B 100 mg/kg; Cyclophosphane 50: cytotoxic molecule 50 mg/kg.*

However, AOS A and B showed no cytotoxic activity in vitro against murine tumor cells of the tsFT201 line. As before, this result suggests that the mechanism of action of sulfated AOS involves an indirect pathway, more specifically that of immunomodulation.

This hypothesis is confirmed by the study of changes in spleen mass as a function of the treatment administered, the spleen being an organ for the production of lymphocyte cells involved in the immune process.

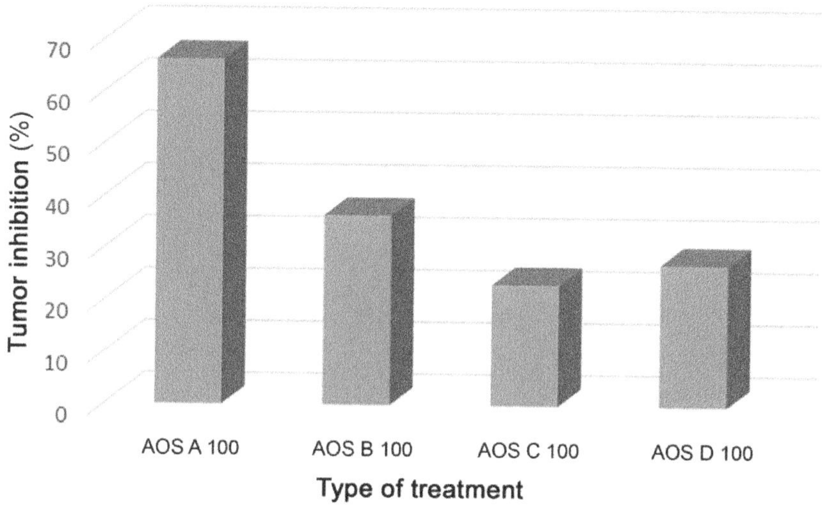

Figure 3.6. *Comparative effect of sulfated or non-sulfated alginate oligosaccharides on inhibition of the Sarcoma 180 tumor inoculated into a murine model (according to Hu et al. (2004))*

COMMENT ON FIGURE 3.6 – *AOS A 100: alginate oligosaccharide A 100 mg/kg; AOS B 100: alginate oligosaccharide B 100 mg/kg; AOS C 100: unsulfated alginate oligosaccharide C 100 mg/kg; AOS D 100: unsulfated alginate oligosaccharide D 100 mg/kg.*

The administration of sulphated AOS (AOS A, AOS B) significantly increases the spleen mass of treated animals compared with controls (see Figure 3.7). For oligosaccharide A, a dose effect was observed, but this did not appear to be the case for AOS B. Indeed, the maximum weight impact on the spleen was obtained for AOS B with an administered concentration of 50 mg/kg, 100 mg/kg or 200 mg/kg (see Figure 3.7).

The results of this and other studies suggest that alginates and their derivatives modulate the immune response of organisms to the cancerization process. Their antitumor action therefore appears to be mainly indirect. The particular physicochemical properties (high molecular weight and viscosity) of native alginates are a hindrance to their formulation and administration as drugs. In this context, the use of oligosaccharides of low molecular weight, and LVAs, is an interesting avenue for adding value to these molecules in the anticancer therapeutic field.

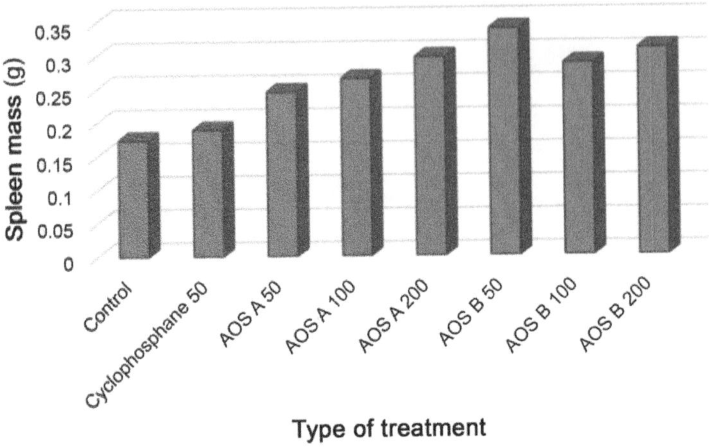

Figure 3.7. *Evolution of spleen mass in mice as a function of treatment administered (according to Hu et al. (2004))*

COMMENT ON FIGURE 3.7. – *AOS A 50: alginate oligosaccharide A 50 mg/kg; AOS A 100: alginate oligosaccharide A 100 mg/kg; AOS A 200: alginate oligosaccharide A 200 mg/kg; AOS B 50: alginate oligosaccharide B 50 mg/kg; AOS B 100: alginate oligosaccharide B 100 mg/kg; AOS B 200: alginate oligosaccharide B 200 mg/kg; Cyclophosphane 50: cytotoxic molecule 50 mg/kg.*

In addition to this potential therapeutic activity, alginates are already used by the pharmaceutical industry for the formulation of numerous drugs (capsules, pills, syrups, gels, suppositories) or for their therapeutic properties (see Table 3.2) (Szekalska et al. 2016). They are also used as biomaterials in bone repair surgery and dentistry.

As in the case of antitumor activities, current research is also focusing on the biological activities of AOS. The activities detected could prove useful in the treatment of severe pathologies such as Alzheimer's disease or HIV infection (see Table 3.3).

Alginates and their derivatives are not the only polysaccharides derived from brown algae with pharmacological activity. Laminarin and fucans, carbohydrate polymers, present in many species of Phaeophyceae, are also molecules of interest for medical research and the promotion of algae in the pharmaceutical field.

Drug	Type of administration	Therapeutic indication
Gastrotuss (Mg alginate)	Oral	Coughing, dysphagia
Algicid (Na alginate)	Oral	Belching
Gaviscon (Na alginate)	Oral	Gastric reflux
Algivon (Ca alginate)	Dermatological	Exudate fixation, wound healing
Algosteril (Ca alginate)	Dermatological	Hemostasis, wound healing
Coalgan (Ca alginate)	Dermatological	Hemostasis, wound healing
Natalside (Na alginate)	Rectal (suppository)	Hemostasis (hemorrhoids)
Progenix (Na alginate)	Periodontal	Tooth regeneration
ChondroArt 3D injection (alginate-agarose)	Arthroscopic	Increased cartilage production and growth

Table 3.2. *Examples of alginate-based pharmaceutical products (according to Szekalska et al. (2016[1]))*

Molecules or product	Biological activity	Therapeutic application
OligoG (poly-G AOS)	Regulation of mucus viscosity	Treatment of pulmonary cystic fibrosis
Heparinoid 911 (poly-M/poly-G sulfated AOS)	Inhibition of DNA polymerase (hepatitis B virus) Inhibition of reverse transcriptase (HIV virus)	Antiviral treatment (HIV, hepatitis B)
HS 971 (poly-M AOS)	Inhibition of the effects of neuroinflammation	Alzheimer's disease and neurodegenerative disorders
Propylene glycol sulfated sodium alginate (PSS AOS)	Thrombin inhibition	Anticoagulant and antithrombotic treatment

Table 3.3. *Examples of reported biological activities for alginate oligosaccharides (according to Szekalska et al. (2016))*

1 See: https://www.vidal.fr/.

3.1.2. *Laminarin*

Laminarin, or laminaran, is not, unlike alginates, a cell wall polymer. It is a reserve polymer initially found in brown algae of the genus *Laminaria,* hence its name. Depending on the season, laminarin content can reach up to 35% of dry matter in *Laminaria digitata*.

In other kelp species, such as *Laminaria saccharina* (*Saccharina latissima*) and *Laminaria hyperborea*, the maximum laminarin content is around 33% of the plant's dry weight (Kadam et al. 2015). Structurally, this polysaccharide consists of 25 to 50 glucose monomers linked by β 1-3 glycosidic bonds (see Figure 3.8).

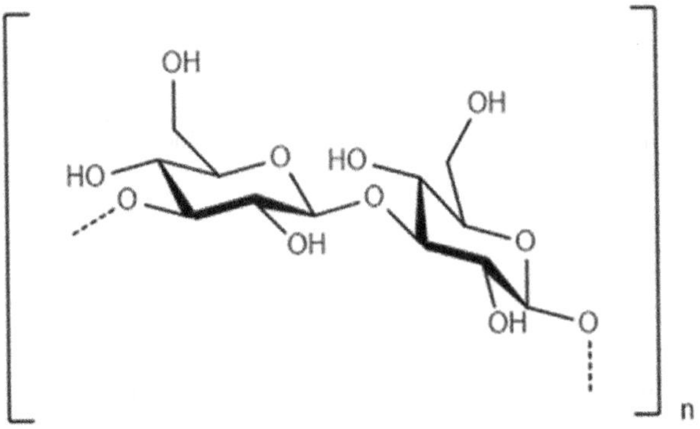

Figure 3.8. *Structure of the laminarin molecule of* Ascophyllum nodosum *(according to Fleurence (2022))*

Laminarin is associated with numerous activities of therapeutic interest (see Table 3.4). In particular, it has a regulatory activity on the immune response and, like alginates, plays an immunomodulatory role (Gupta and Abu-Ghanam 2011).

The mechanism involved concerns the viability of immune cells or thymocytes produced by the thymus. Laminarin and its enzyme degradation product, laminarin oligosaccharide, limit thymocyte apoptosis in vitro (Kim et al. 2006) (see Figure 3.9). As such, laminarin appears to be a regulator of the immune system and inflammatory response.

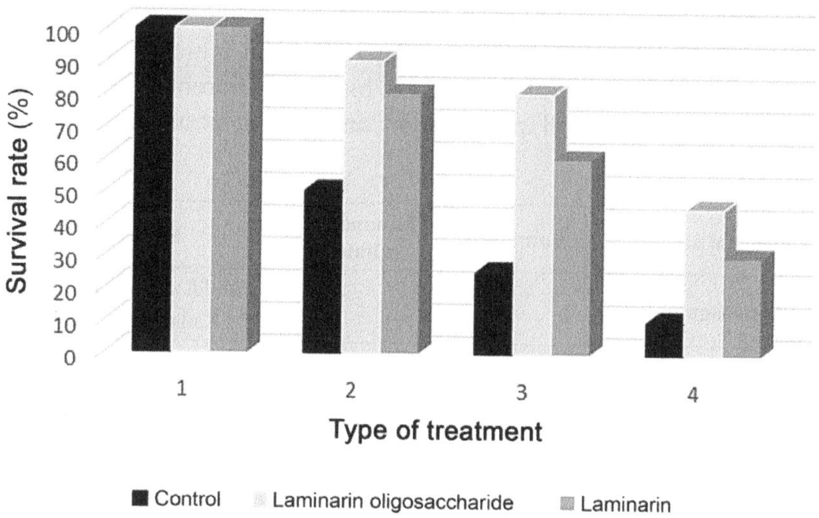

Figure 3.9. *Impact of in vitro administration on thymocyte cultures of laminarin or laminarin oligosaccharide on cell survival over time (1: 0 days; 2: 3 days; 3: 6 days; 4: 12 days) (according to Kim et al. (2006))*

In addition to this type of activity, laminarin is described in the literature for its antitumoral, anti-inflammatory, antioxidant or antiviral activities (see Table 3.4). Its sulfated derivative obtained by chemical modification is notably endowed with anticoagulant activities.

In addition to the various biological activities mentioned above, laminarin is also known for its direct antitumoral action on certain malignant cell lines (Zargarzadeh et al. 2020). In particular, it acts by triggering apoptosis in tumor cells. This laminarin-induced cell death mechanism has been described for the LoVo (colorectal cancer), Y79 (retinoblastoma), A-375 (melanoma) and HepG2 (liver cancer) cell lines.

The chemical modification of laminarin, and more specifically its sulfation, significantly increases the polysaccharide's anticancer activity. Incubation of LoVo line cells in the presence of sulfated laminarin at a concentration of 1,600 µg/mL resulted in a decrease in cell population proliferation of almost 86%. This decrease in cell proliferation was only 38% with native, i.e. unsulfated, laminarin.

Laminarin is also thought to be involved in the inhibition of angiogenesis, the process that leads to tumor vascularization and initiates metastasis. In particular, the polysaccharide inhibits the binding of fibroblast growth factor, a factor that induces cell proliferation and angiogenesis (Perumal et al. 2023).

Biological activity	Laminarin	Laminarin sulfated	References
Anticoagulant	-	+	Holdt and Kraan (2011) Kadam et al. (2015)
Antioxidant	+	nd	Kadam et al. (2015)
Anti-inflammatory	+	nd	Neyrinck et al. (2007) Kadam et al. (2015)
Antiallergic	nd	nd	Karuppusamy et al. (2022)
Immunomodulator	+	nd	Gupta and Abu-Ghanam (2011) Kadam et al. (2015)
Immune cell stimulation (B and T lymphocytes)	+	nd	Holdt and Kraan (2011)
Antitumor	+	+	Zagarzadeh et al. (2020) Jin et al. (2022)
Antiangiogenesis	+	+	Perumal et al. (2023)
Hypotensive	+		Holdt and Kraan (2011)
Antilipid hypocholesterolemic	+	+	Mookerjea and Hawkins (1958) Holdt and Kraan (2011)
Antiviral	+	nd	Lomartire and Goncalves (2022)

Table 3.4. *Some examples of biological activities of therapeutic interest associated with laminarin or its sulfated derivative (nd: not described in the literature referenced in Table 3.4)*

Another laminarin derivative, oxidized lamarin (OLA), or glucoglucuran, is also described for its anticancer activities. In this modified polymer, some of the glucose monomers are oxidized to glucuronic acid (see Figure 3.10). OLA has been shown to be particularly effective in limiting the proliferation of A549 lung cancer cells in vitro. Depending on the dose administered, oxidized laminarin causes varying degrees of mortality in cancer cells after

12 hours (see Figure 3.11). The necessary concentration of OLA for 50% cell death is estimated at 2.70 mg/mL.

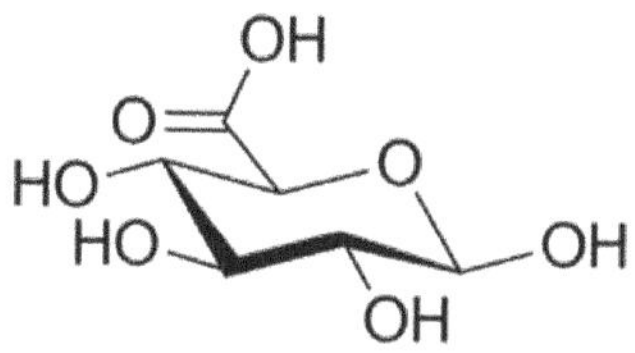

Figure 3.10. *Structure of glucuronic acid derived from the oxidation of glucose and constituent of modified laminarin (according to Perumal et al. (2023[2]))*

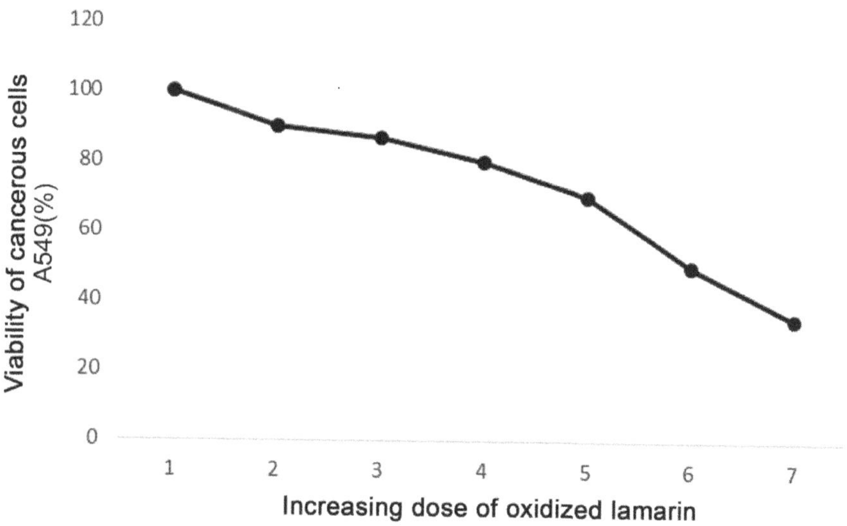

Figure 3.11. *Effect of in vitro administration of increasing doses of laminarin on the viability of A549 lung cancer cells (1: 0 mg/mL; 2: 0.5 mg/mL; 3: 1 mg/mL; 4: 1.5 mg/mL; 5: 2 mg/mL; 6: 2.5 mg/mL; 7: 3 mg/mL) (according to Jin et al. (2020))*

Laminarina polysaccharide found in brown algae shows that, like alginates, it has a range of biological activities of interest to human health. Its particularly high content in Laminaria and its ease of extraction are

2 See: https://commons.wikimedia.org/w/index.php?curid=8545380/.

undeniable assets for the development of this polymer as a new pharmaceutical active ingredient.

Its integration into anticancer therapeutic strategies appears to be a potential avenue for the development of new drugs. Its effect on the immune system and its direct action on malignant cell lines offer clear advantages for medical research.

This activity seems, moreover, to be enhanced by the chemical modification of laminarin, in particular its degree of sulfation. The presence of sulfate groups on algal polysaccharides is often decisive for their biological activity. This is particularly true of fucans, another highly sulfated polysaccharide found in brown algae.

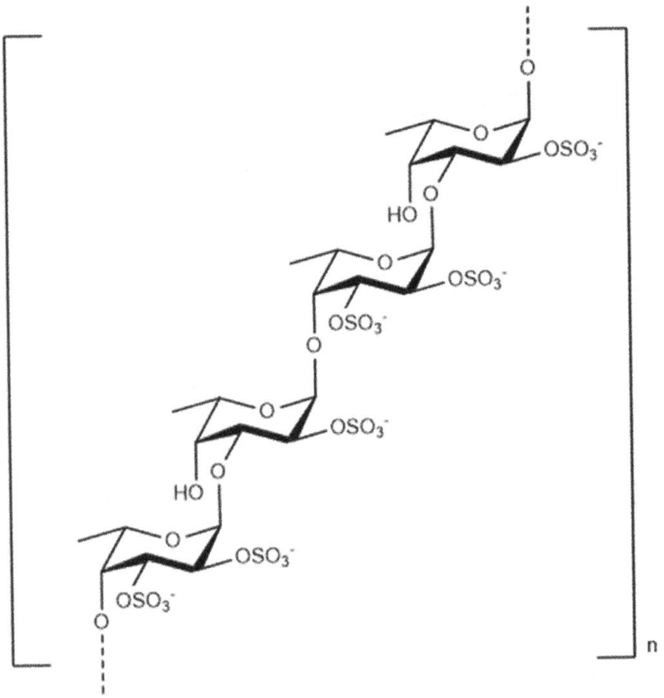

Figure 3.12. *Structure of a homofucan, or fucoidan, from brown algae (*Ascophyllum nodosum*) (according to Fleurence (2018))*

3.1.3. *Fucans*

Fucans are sulfated polysaccharides present in brown algae. They are classified into three distinct families according to their chemical composition. The fucoidans or homofucans (see Figure 3.12), xylofucoglucuronans known as ascophyllans and glucuronofucoglycans or sargassans (Fisher et al. 1991). Fucoidans are distinguished from other fucans by their fucose content. Fucoses account for 50–90% of the polymer's constituent carbohydrates. By comparison, ascophyllans and sargassans contain 25% and 50% fucose respectively. From a cellular point of view, fucans are distributed between the intercellular space and the cell wall (Kloareg 1991).

Biological activities of therapeutic interest are often associated with fucoidans or homofucans. These mainly concern anticoagulant or antithrombotic activity. However, other therapeutic properties (anticancer, anti-inflammatory, etc.) have also been reported for fucoidans and other fucans (see Table 3.5) (Wijesinghe and Jeon 2012).

The anticoagulant activity of fucoidans has been known since 1957 (Springer et al. 1957). At that time, it was demonstrated on the fucoidan fraction of the brown alga *Fucus vesiculosus*. This type of activity is also described for another fucal, *Pelvetia canaliculata* (Colliec et al. 1991). In this species, the fucoidan fraction is active in inhibiting thrombin production. It inhibits thrombin production by almost 40% at a concentration of 30 µg/mL. This level of inhibition is also achieved with heparin, but at a concentration of 0.3 µg/mL.

Anticoagulant activity is not only associated with fucoidans; other fucans also appear to possess this type of activity (Fischer et al. 1991). The anticoagulant and antithrombotic activities of fucans appear to be closely linked to their degree of sulfation (see Figure 3.13).

But this criterion is not the only one involved. The intracellular origin of fucans and therefore their nature are also an important factor. Depending on the algal species and extraction techniques, the nature of the fucans obtained varies (parietal fucans, intercellular fucans). The antithrombotic activities of these different fucan fractions vary considerably, as shown in Figure 3.14.

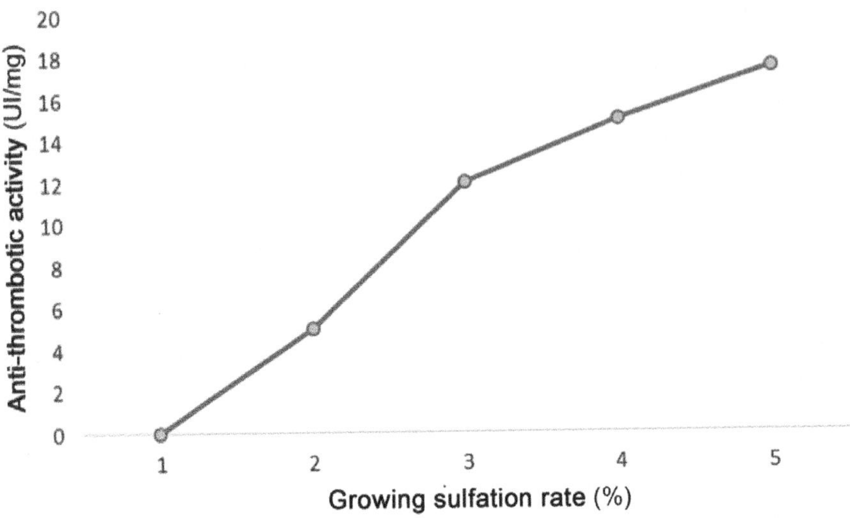

Figure 3.13. *Impact of the sulfation rate of fucans of* Fucus vesiculosus *on in vitro antithrombotic activity (according to Fischer et al. (1991)) (1: 0% sulfation; 2: 10% sulfation; 3: 20% sulfation; 4: 30% sulfation; 5: 40% sulfation)*

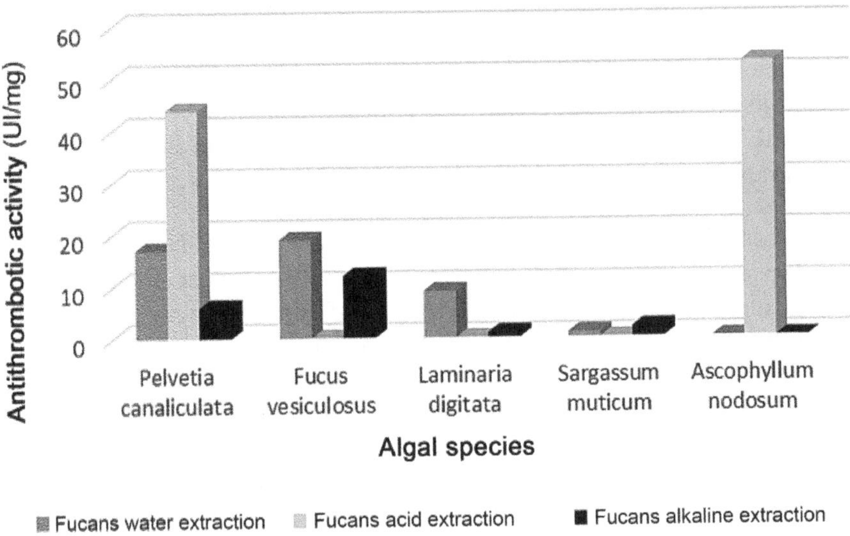

Figure 3.14. *Antithrombotic activity of fucans obtained by different extraction methods from different species of brown algae (according to Fischer et al. (1991))*

Algal species	Anticoagulant, antithrombotic	Immunomodulation	Anti-inflammatory	Antitumor	Antiviral	Neuroprotective
Ecklonia cava	+	nd	+	+	nd	nd
Fucus evanescens	+	nd	nd	+	nd	nd
Ascophyllum nodosum	+	nd	nd	nd	nd	nd
Hizikia fusiformis	+	nd	nd	nd	nd	nd
Fucus vesiculosus	nd	+	nd	nd	+	nd
Undaria pinnatifida	nd	nd	nd	+	+	nd
Laminaria japonica	nd	nd	+	nd	nd	+
Sargassum fulvellum	+	nd	nd	nd	nd	nd
Laminaria cichorioides	+	nd	nd	nd	nd	nd

Table 3.5. *Examples of therapeutic activities of interest associated with fucans extracted from different species of brown algae (according to Wang et al. (2012); Wijesinghe and Jeon (2012)) (nd: not described by Wijesinghe and Jeon (2012))*

Fucans and their subfamily, the fucoidans, like many algal polysaccharides, are also associated with activities of interest to human health (see Table 3.5). Nevertheless, these sulfated polysaccharides are still mainly described in the literature for their anticoagulant or antithrombotic activities.

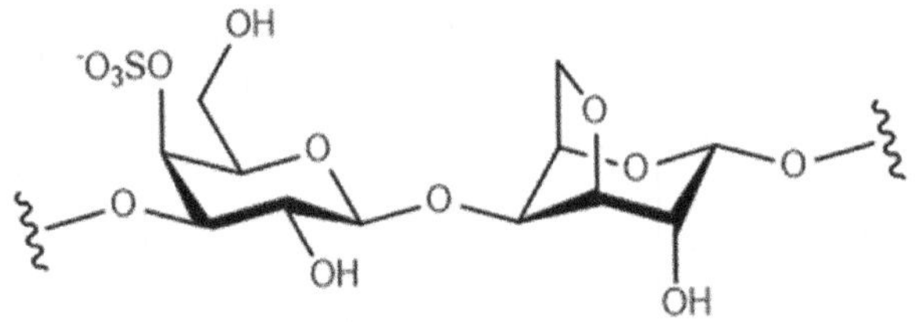

a) D-galactose 4 sulfate   3, 6-anhydro-D-galactose

κ-carrageenan

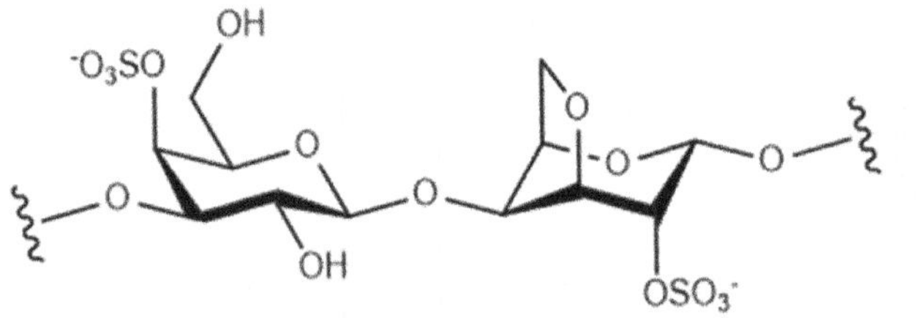

b) D-galactose 4 sulfate   3, 6-anhydro-D-galactose 2 sulfate

ι-carrageenan

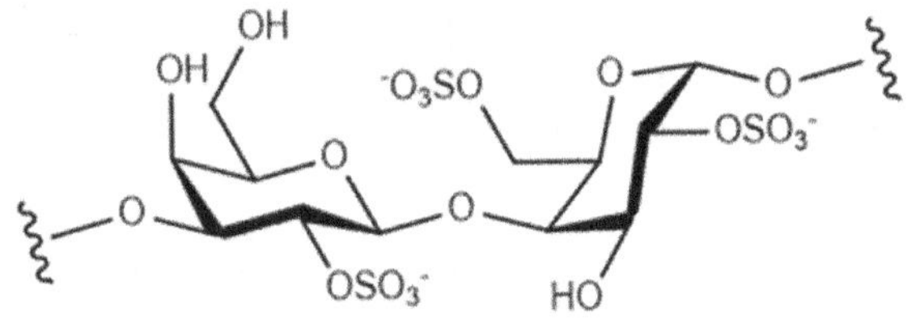

c) D-galactose 2 sulfate   D-galactose 2,6 sulfate

λ-carrageenan

Figure 3.15. *Chemical structure of the three carrageenan families (according to Fleurence (2018))*

3.1.4. *Carrageenans*

The term "carrageenan" derives from the common name of the red alga *Chondrus crispus*, which is called carrageen moss in the UK and carrageen in Ireland (Necas and Bartosikova 2013). Carrageenans are sulfated polysaccharides present in many species of red algae belonging to the genera *Chondrus*, *Eucheuma*, *Kappaphycus* and *Mastocarpus* (Perez 1997). The carrageenan family is divided into three main subgroups with different structural and physico-chemical characteristics (see Figure 3.15 and Table 3.6). Their high molecular weight and gelling or texturizing capacity are the main physico-chemical properties associated with these polysaccharides.

The three main subfamilies of carrageenans are κ-carrageenans, ι-carrageenans and ʎ-carrageenans respectively.

κ-carrageenans are polymers derived from the chain of a D-galactose-4-sulfate molecule with a 3,6-anhydro-D-galactose molecule. ι-carrageenans are obtained via the repetition of a disaccharide composed of a molecule of D-galactose-4-sulfate and a molecule of 3,6 anhydro-D-galactose-2-sulfate. ʎ carrageenans, on the contrary, lack an anhydrous bond. They are formed by repeating the chain of a D-galactose-2-sulfate molecule with a D-galactose-2,6-sulfate molecule (see Figure 3.15).

Other categories of carrageenan, less frequently described in the literature and therefore less studied, are nevertheless present. These are the ε and μ carrageenans (Necas and Bartosikova 2013). All these carrageenan subfamilies (κ, ι, ʎ, ε, μ) possess 22–35% sulfate groups and have molecular weights well in excess of 100 kDa (see Table 3.6). Natural carrageenans have particularly high molecular weights, ranging from 1.5×10^6 to 2×10^7 Da (Necas and Bartosikova 2013). Carrageenans used as industrial additives under the name E 407 have lower molecular mass values, generally ranging from 1.10^5 to 8.10^5 Da (see Table 3.6). Carrageenans can also be depolymerized. These chemically or enzymatically obtained oligo-carrageenans have a lower molecular weight than the original polysaccharide.

Type of carrageenan	Molecular mass (Da)	Physicochemical properties	Main source
κ-carrageenan	738,000	Gelling agent	*Chondrus crispus* *Eucheuma spinosum* *Eucheuma cottonii*
ι-carrageenans	440,000	Gelling agent	*Eucheuma spinosum* *Eucheuma cottonii*
λ-carrageenans	700,000	Texturizer	*Chondrus crispus*

Table 3.6. *Examples of characteristics of different types of carrageenans (according to Viebke et al. (1995); Almutairi et al. (2013))*

Carrageenans are associated with numerous properties of therapeutic interest. The main ones concern antiviral and antitumoral activities. However, these polymers have also been described for their immunosuppressive activity (Thomson and Fowler 1981). Several immune response mechanisms are involved in this carrageenan action. These include a reduction in the response to T-dependent antigens, and hence reduced activation of B lymphocytes, inhibition of the complement response, a delay in the hypersensitive response and impaired cytotoxic reactions. They can also lower the hepatic phagocytic response, which is part of the defense mechanism.

In fact, injection of carrageenans to mice previously inoculated with sheep red blood cells had a negative effect on their elimination by hepatic phagocytosis (see Figure 3.16). This last point suggests that a partial inhibition by algal polysaccharides of cell of the Kupffer cells responsible for eliminating used red blood cells and other foreign or damaged cells from the liver.

Independently of this immunosuppressive activity, carrageenans are also known to be toxic when administered intravenously. The dose required to achieve 50% lethality (LD50) is particularly low compared with that required for oral administration (see Table 3.7). Indeed, the oral dose required to trigger toxicity ranges from 2,640 to 9,140 mg/kg body weight, depending on the animal models studied. In contrast, carrageenans administered by intravenous injection can induce toxicity at doses of less than 50 or even 1 mg/kg body weight. This last point shows that the administration of carrageenans by injection for therapeutic purposes is a pathway that should be avoided in the event of drug development. More generally, carrageenans, especially when purified, show moderate toxicity when administered orally, by inhalation or by dermal application (Weiner 1991).

Type of carrageenan	Species	Type of administration	LD50 (mg/kg body weight)
Kappa/lambda carrageenans	Mouse	Oral	9,150
Kappa/lambda carrageenans	Rabbit	Oral	2,640
Kappa/lambda carrageenans	Rabbit	Intravenous	1–20 (LD 100)
Kappa/lambda carrageenans	Guinea pig	Intravenous	> 10
Lambda carrageenans	Guinea pig	Intravenous	< 1
Carrageenans	Rabbit	Intravenous	< 50
Iota carrageenan	Rat	Oral	> 5,000
Iota carrageenan	Rat	Inhalation	> 930

Table 3.7. *Acute toxicity of carrageenans according to mode of administration (oral, intravenous, inhalation) (according to Necas and Bartosikova (2013))*

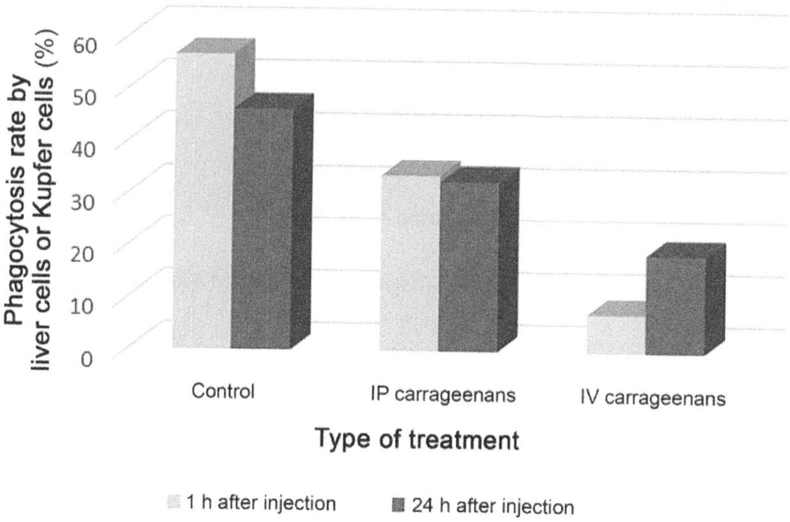

Figure 3.16. *Impact of carrageenan injection by intraperitoneal (ip) or intravenous (iv) injection on the phagocytic activity of hepatic Kupffer cells (according to Thomson and Fawler (1981))*

In addition to the above-mentioned negative effects, carrageenans also have interesting human health properties. These in particular include antiviral activities. A query on the academic search engine Google Scholar brings up 27,300 results when the keywords *carrageenan-antiviral* are used.

Carrageenans are described as inhibitors of enveloped viruses, some of which are responsible for human pathologies (AIDS virus, herpes simplex virus (HSV), cytomegalovirus, rhinovirus) (Necas and Bartosikova 2013; Lomartine and Goncalves 2022). Several mechanisms of action are described in the literature. They range from direct inactivation of the virus to preventing the completion of the various stages of the viral cycle (see Figure 3.17).

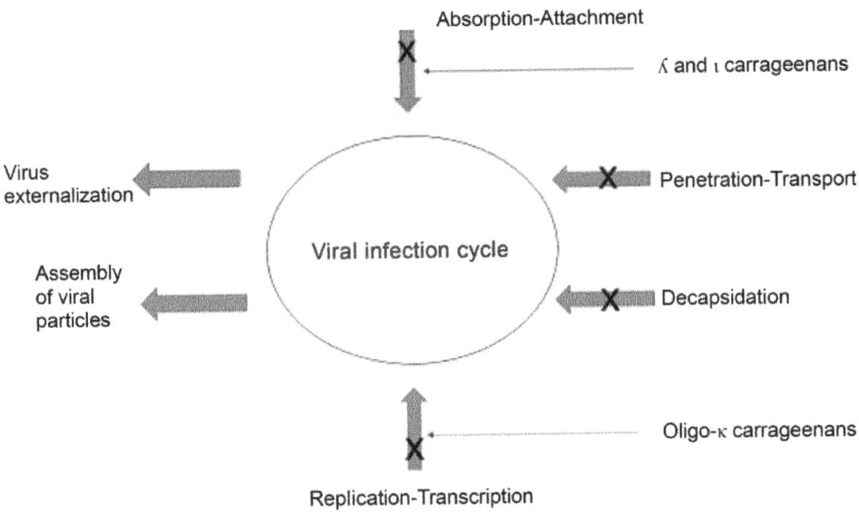

Figure 3.17. *Classic viral infection cycle*

Carrageenans, like sulfated polysaccharides, have the ability to bind to viral envelopes, thereby limiting their capacity for absorption or attachment onto the cell membrane. This mechanism, which applies directly to the virus, reducing its capacity for infection and thus reproduction, is considered to be virucidal activity. With regard to the HSV, the molecular mechanism of action of λ-type carrageenans is well described. Carrageenans are thought to induce structural changes in the virus gB and gC glycoproteins, which are necessary for fusion of the viral envelope with the cell membrane. The virucidal activity of carrageenans is closely linked to their molecular weight, their degree of sulfation and the distribution of sulfate groups on the carbon skeleton. Up to a mass of 100 kDa, the virucidal activity of carrageenans would increase, only to decrease beyond this mass value (Wang et al. 2012). Negatively charged carrageenans can also interact with positively charged cell membrane receptors. When this type of interaction occurs, the virus binding sites on the cell membrane are masked and no longer accessible. The initial stage of infection, i.e. virus absorption or attachment, is no longer possible (see Figure 3.17). In particular, κ- and ι-carrageenans have been described as potent inhibitors of the attachment of herpes simplex viruses (HSV 1, HSV 2) (Wang et al. 2012).

Carrageenans may also limit or even inhibit other stages of the viral cycle. Jointly supplied ʎ and ι carrageenans are likely to inhibit the penetration stage of dengue virus (DENV), and especially its release from endosomes responsible for transporting viral particles from membrane to cytoplasm. In the same viral model, ι-carrageenans interfere with the decapsidation process, an essential step in the cell's synthesis of new viruses (transcription-replication). Polysaccharides interact with viral envelope glycoproteins (gE), inhibiting the decapsidation process.

ι-carrageenans are also described for their ability to strongly interfere with the early stages of viral replication in rhinovirus (HRV), thereby inhibiting virus multiplication (Wang et al. 2012).

Like replication, transcription is a key step in the viral cycle. It represents one of the last stages in the virus reproductive cycle before assembly and externalization of viral particles. Certain low molecular weight derivatives, such as carrageenan oligosaccharides, have been shown to inhibit Influenza A virus (IAV) transcription (Wang et al. 2011). The κ-carrageenan oligosaccharide with a molecular mass of 2 kDa is highly effective in inhibiting the reproductive process of the virus. The effective dose to induce 50% inhibition of viral reproduction is 32.1 μg/mL for this oligosaccharide, whereas it is between 8 and 16 times higher for oligosaccharides with molecular masses of 3 and 5 kDa (see Figure 3.18). It is almost as effective as Ribavirin, a classically used antiviral drug.

Despite the interest in the antiviral activities of carrageenans,, these high molecular weight sulfated polysaccharides have a high molecular weight, high viscosity and low penetration capacity. Added to this is the toxicity reported during intravenous injection. Oligo-carrageenans, with their lower molecular weight and lower viscosity, appear to be better candidates for therapeutic application in human health. The depolymerization process for ʎ- and κ-carrageenans, which is easily performed, relies on ultrasonication, at room temperature and in the presence of ascorbic acid (Yamada et al. 2000). Depolymerized polysaccharides, such as carrageenans, show antiviral activities of interest (see Table 3.8).

Carrageenans and their depolymerized forms have been described for their anti-HIV activity (Yamada et al. 1997). This activity differs according to the type of carrageenan. The gelling capacity of κ- and ι-carrageenans under the operating conditions set up by Yamada prevents any evaluation of anti-HIV activity. This last point is important, as it highlights the difficulty of assessing biological activities using sulfated polysaccharides generating high viscosity. Subsequent studies on the antiviral activities of κ- and ι-carrageenans have taken this experimental limitation into account and adapted their experimental conditions. On the contrary, ʎ-carrageenans with texturizing rather than gel-forming properties proved easier to use for this type of study.

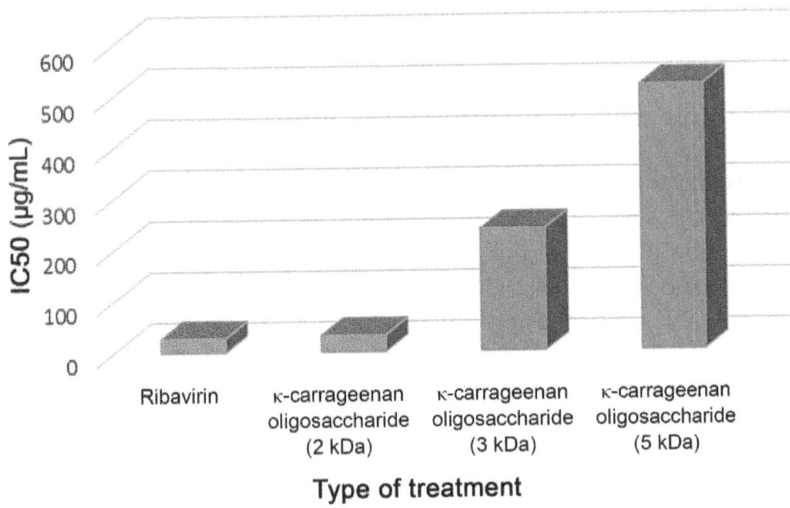

Figure 3.18. *Effect of κ-carrageenan oligosaccharides of different molecular weights on 50% inhibition of IAV virus production (according to Wang et al. (2011))*

Type of sulfated polysaccharides	Antiviral activity
Carrageenans	HVS HIV HPV IAV DENV
Oligo-carrageenans	HIV IA RABV

Table 3.8. *Examples of antiviral activities associated with carrageenans or oligo-carrageenans (HSV: herpes simplex virus; HIV: human acquired immunodeficiency virus; HPV: human papillomavirus; IAV: influenza A virus; DENV: dengue virus; RABV: rabies virus) (according to Wang et al. (2012) and Alvarez-Vinas et al. (2021))*

λ-carrageenans are particularly effective at inhibiting the HIV virus. The minimum concentration for 100% inhibition of viral particles is established at 7.8 μg/mL of polysaccharides (see Figure 3.19). The depolymerized forms, λ-carrageenan oligo 1 and λ-carrageenan oligo 2, have an IC_{100} of 3.9 μg/mL, i.e. two times higher anti-HIV activity than the initial polymerized form.

Molecular weight reduction appears to be an important factor in the anti-HIV activity of λ-carrageenans. Type I and II oligo-carrageenans, with molecular weights of 280 kDa and 46 kDa respectively, are indeed more active than the initial polymer, with a mass of 1,700 kDa.

However, this is not the only criterion involved. The number of sulfate groups per disaccharide also appears to be important for anti-HIV activity. λ-carrageenan oligo 3, characterized by its low molecular mass (7 kDa), proved to be the least active, with an IC_{100} of 250 μg/mL, compared with the initial polymer and the other oligo-carrageenans (see Figure 3.19).

This result suggests that the anti-HIV activity of λ-carrageenans and its oligosaccharide derivatives are linked to a minimum carbon chain length and a minimum number of sulfate groups (> 2) (see Figure 3.20).

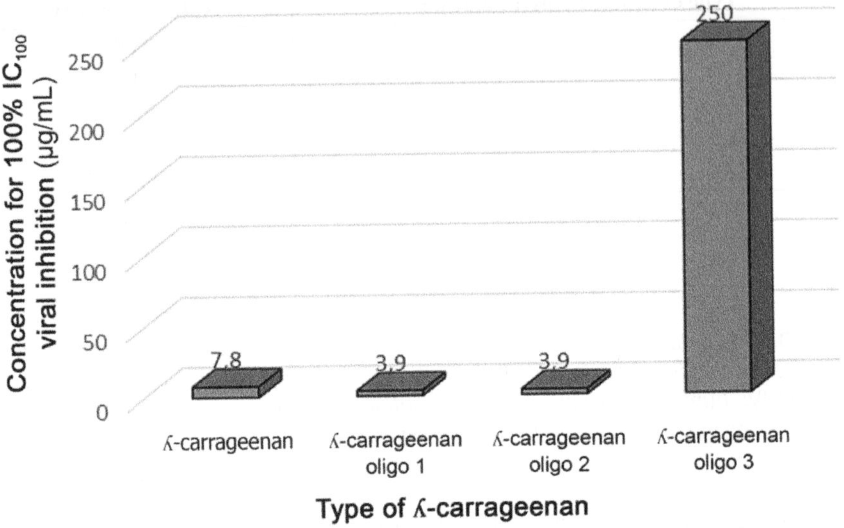

Figure 3.19. *Anti-HIV activity of ʎ-carrageenan and its oligo-carrageenans (ʎ-carrageenan: 1,700,000 Da; ʎ-carrageenan oligo 1: 280 kDa; ʎ-carrageenan oligo 2: 46 kDa; ʎ-carrageenan oligo 3: 7 kDa) (according to Yamada et al. (1997))*

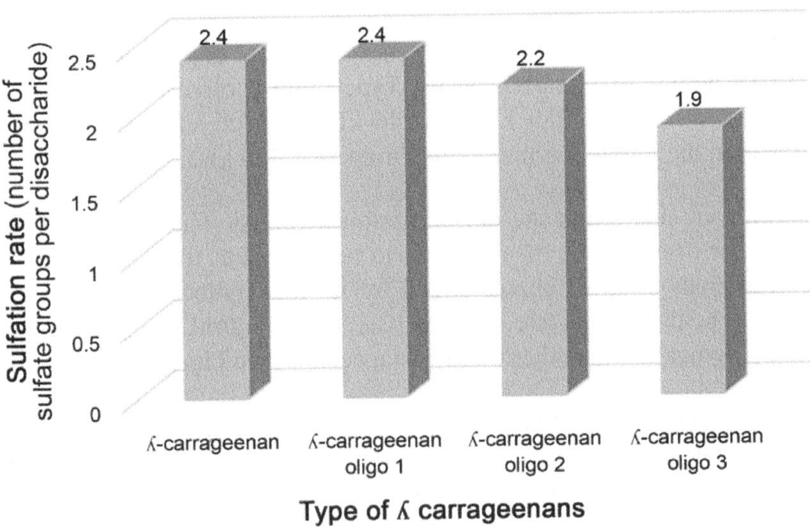

Figure 3.20. *Sulfation rates of the different types of ʎ-carrageenans tested for anti-HIV activity (according to Yamada et al. (1997))*

Generally speaking, the most active forms in terms of antiviral activity are, in descending order, λ-, ι- and κ-carrageenans (Alvarez-Vinas et al. 2021). The degree of sulfation and the position of sulfate groups also play an important role in the activity of undenatured carrageenans.

ι-and κ-carrageenans, of commercial origin, and therefore derived from an industrial process, have masses below 10^6 Da, significantly lower than the molecular masses reported for native carrageenans. Numerous studies on antiviral activities are often carried out using this type of food-grade polysaccharide. The fact remains, however, that these carrageenans, with their gelling properties, are used at relatively low concentrations.

Today's strategy for investigating antiviral activities is based primarily on depolymerized carrageenans whatever the origin of the initial polysaccharide (λ-, ι- and κ-carrageenans).

As with native carrageenans, the degree, or rate, of sulfation plays an important role in the antiviral activity of carrageenan oligosaccharides. Already reported for λ-carrageenan oligosaccharides (see Figure 3.20), this is also verified for κ-carrageenan oligosaccharides (see Figure 3.21).

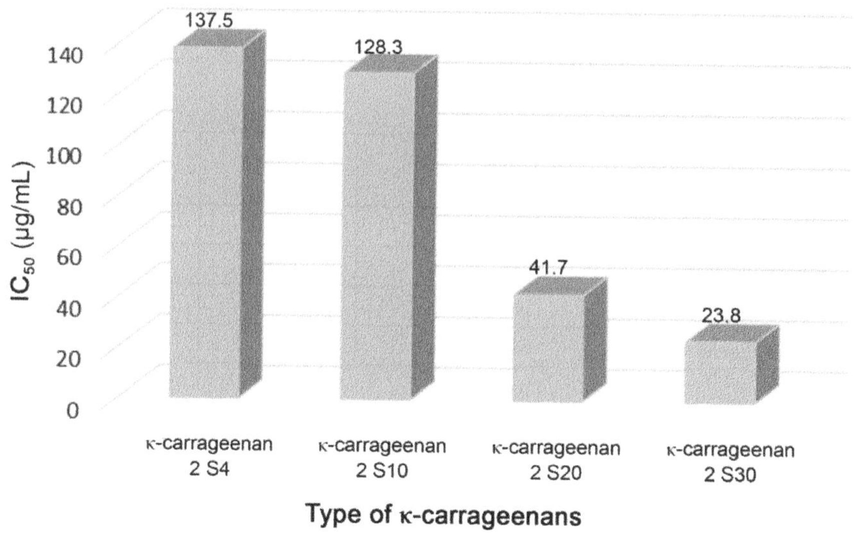

Figure 3.21. *Effect of κ-carrageenan oligosaccharides on the inhibition of 50% of particles, S4 (4%), S10 (10%), S20 (20%) and S30 (30%) sulfation (according to Alvarez-Vinas et al. (2021))*

The antiviral activity of κ-carrageenan oligosaccharides, of low molecular mass (2 kDa) against influenza A virus (influenza A), seems to be directly linked to their degree of sulfation (see Figure 3.21). The minimum concentration for reproduction inhibition of 50% of viral particles is high for a degree of sulfation of 4% (IC_{50}: 137.5 µg/mL), indicating low antiviral activity. In contrast, the IC_{50} is only 23.8 µg/mL, with an oligosaccharide whose degree of sulfation is 30%, i.e. almost six times higher antiviral activity than before (see Figure 3.21).

Carrageenans, sulfated polysaccharides, such as fucoidans, possess antitumoral properties, as often demonstrated in vitro on malignant cell lines. As in the previous case, these activities are associated with native polymers, and also with their oligosaccharide derivatives. The effects of carrageenans differ greatly depending on the type of carrageenan, the concentration of use and the type of tumor targeted. For example, the commercially available κ-carrageenan has no antiproliferative activity on the tsFT210 mutant breast carcinoma cell line (Khotimchenko et al. 2020). In contrast, the same type of carrageenan was moderately effective against the Hela tumor line (IC_{50} 550.8 µg/mL). ʎ-carrageenan, also from a commercial source, also showed an IC_{50} of 475 µg/mL against this line. κ-carrageenan extracted from the red alga *Kappaphycus alvarezii*, on the contrary, proved effective in inhibiting numerous tumor cell lines in vitro (see Table 3.9). It showed a minimum 50% inhibitory concentration on cell proliferation of 47.85 µg/mL on the MG 63 cell line.

Type of carrageenan	**Cell line**	**IC_{50} (µg/mL)**	**Cancer type**
κ-carrageenan	MCF-7	103.2	Breast cancer
κ-carrageenan	HT 29	73.87	Colon cancer
κ-carrageenan	HepG2	56.71	Liver cancer
κ-carrageenan	MG 63	47.85	Osteosarcoma

Table 3.9. *In vitro inhibitory activity (IC_{50}) of tumor cell proliferation in different cell lines (according to Khotimchenko et al. (2020))*

This result highlights the wide variability in antitumor activities associated with carrageenans or depolymerized carrageenans. The type of carrageenan, whether of commercial origin or directly extracted from the algae, and the type of tumor line are also factors in variability.

Carrageenan oligosaccharides exhibit antitumor activity in vitro, and also in vivo against solid tumors such as Sarcoma 180 (Hu et al. 2006). Depolymerization of commercial gelling and texturizing carrageenans by a bacterial κ-carrageenase enables the production of two types of oligosaccharides (gelling or texturizing) with low molecular weights (681 and 798 Da). The sulfation rate of these oligosaccharides ranges from 17.2 to 21.8%.

These two types of oligosaccharides have been shown to be effective in inhibiting sarcomic tumors. Oligo-carrageenan, with its texturizing properties, has the highest tumor-inhibiting activity, with an inhibition rate of 70.8% when administered at a dose of 100 mg/kg body weight (see Figure 3.22).

The antitumor and antiangiogenesis activity of κ-carrageenan oligosaccharides, obtained by enzymatic hydrolysis, in the context of Sarcoma 180 is also well described in more recent work (Yao et al. 2016).

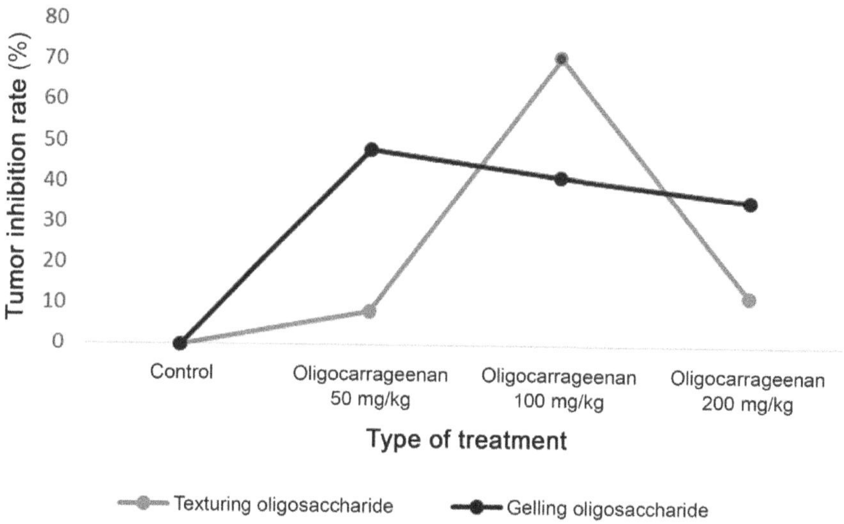

Figure 3.22. *Effect of oligosaccharides supplied at different doses (mg/kg body weight) and derived from enzymatic degradation of carrageenan regarding in vivo inhibition of the Sarcoma 180 tumor in a murine model (according to Hu et al. (2006))*

Enzymatic hydrolysis of κ carrageenan leads to a mixture of three oligosaccharides with molecular masses of 425.2 Da, 1,242.9 Da and 1,681 Da respectively. These oligosaccharides, whose structure is perfectly determined, are respectively κ-neocarrabiose-sulfate, κ-neocarrahexaose-sulfate and κ-neocarraoctaose-sulfate. This mixture, or KOS (kappa oligosaccharides carrageenan), shows sarcoma tumor inhibitory activity that increases with the dose administered to the diseased mouse (mg/kg body mass). The highest antitumor activity was obtained at a dose of 200 mg KOS/kg murine body mass (see Figure 3.23).

The KOS oligosaccharide mixture showed a tumor inhibition rate of 56.8% at a dose of 200 mg/kg body weight. This inhibition rate is comparable to, albeit lower than, that of cyclophosphamide, a potent antitumor agent used in chemotherapy (see Figure 3.24).

The oligosaccharides in the KOS mixture were also active against the MCF-7 cell line responsible for mammary tumors. At a dose of 200 mg/kg body weight administered daily, they limited tumor enlargement in mice inoculated with these malignant cells. Indeed, in this case, daily administration of the oligosaccharide mixture limited tumor volume by a factor of 3, compared with its evolution over 28 days in the absence of treatment (see Figure 3.25).

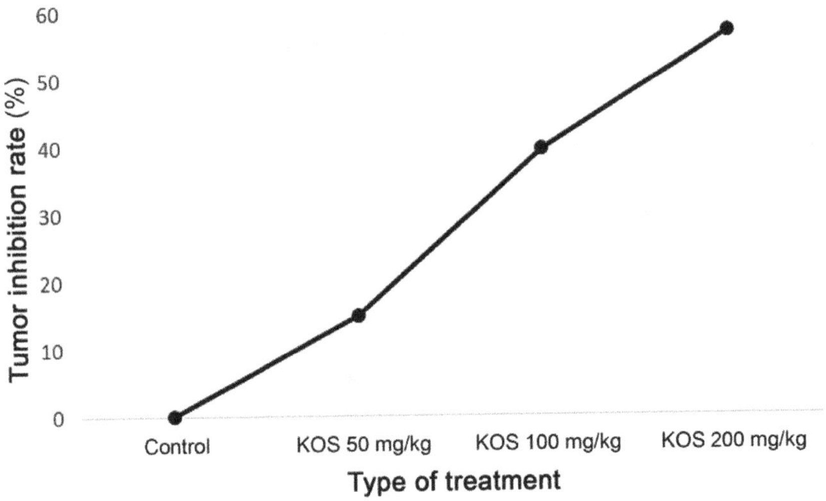

Figure 3.23. *Antitumor activity of a mixture of oligosaccharides of κ-carrageenan (KOS), administered at different doses (mg/kg body weight), on Sarcoma 180 tumor in murine model (according to Yao et al. (2014))*

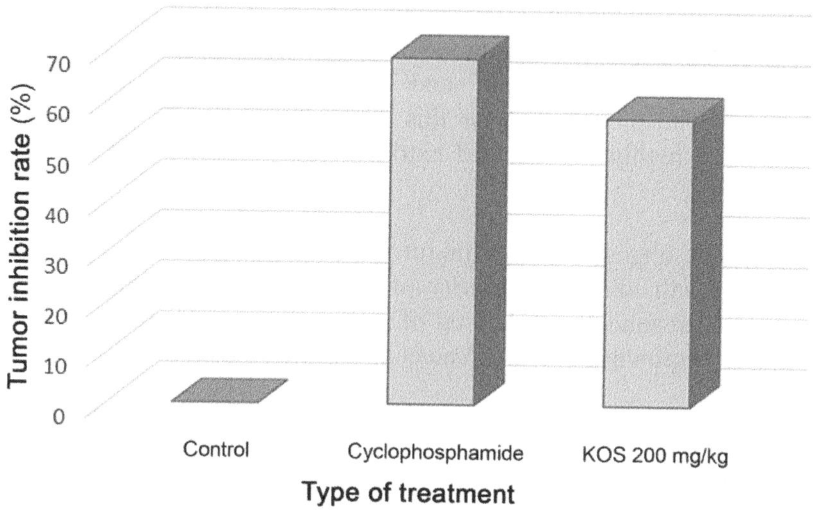

Figure 3.24. *Comparative effect of the antitumor agent Cyclophosphamide and KOS (200 mg/kg) on the tumor inhibition rate of Sarcoma 180 in murine models (according to Yao et al. (2014))*

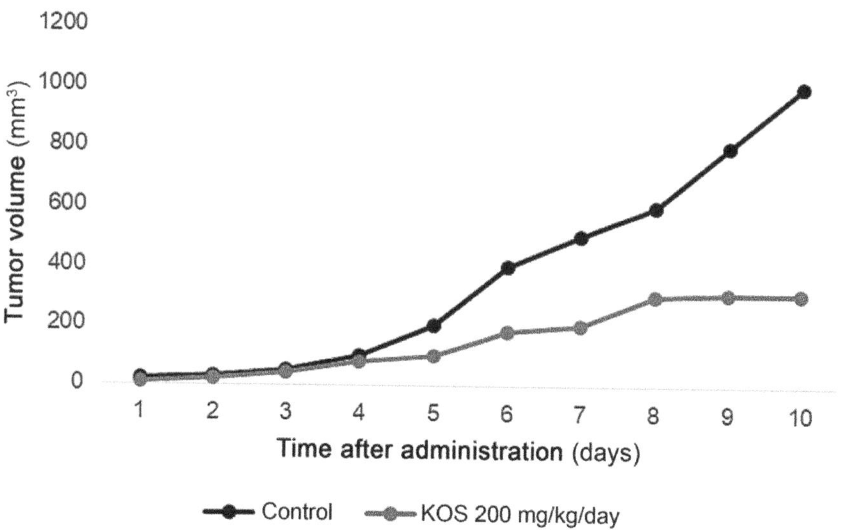

Figure 3.25. *Effect of daily administration of KOS mixture at a dose of 200 mg/kg body mass on MCF-7 tumor volume over time (1: 1st day; 2: 4th day; 3: 7th day; 4: 10th day; 5: 13th day; 6: 16th day; 7: 19th day; 8; 22nd day; 9: 25th day; 10: 29th day) (according to Yao et al. (2014))*

The mixture studied was also found to have an effect on angiogenesis. This process of growth and production of new vessels is an important mechanism in carcinogenesis, and particularly in the production of metastases. The model used for this type of study is the chorioallantoic membrane, a highly vascularized extra-embryonic structure in the chicken amniotic egg.

The KOS oligosaccharide mixture shows, at a dose of 5 µg/egg, a vascular growth inhibitory activity of almost 75% (see Figure 3.26). By comparison, treatment with 500 µg of hydroxycortisone led to an inhibition rate of vessel growth of 78.3% (Yao et al. 2014).

Carrageenans are macromolecules with a high molecular weight and high viscosity. Although endowed with properties of primary interest in human health, such as antiviral activities, these sulfated polysaccharides cannot be injected in their native form. Indeed, their intravenous administration can be lethal. In this context, carrageenan oligosaccharides, with their low molecular weight, low viscosity and biological activities, appear to be much better candidates for therapeutic applications than non-depolymerized carrageenans.

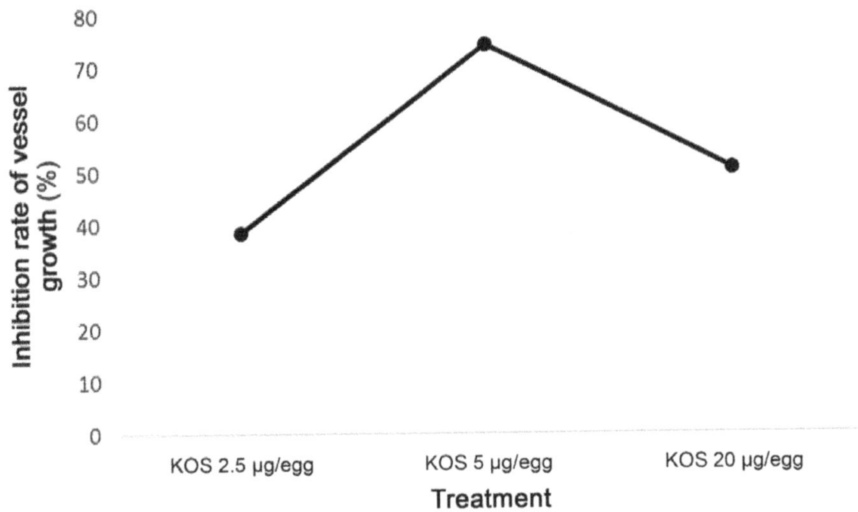

Figure 3.26. *Inhibitory effect on angiogenesis on egg chorioallantoic membrane of different doses of κ-carrageenan oligosaccharides (KOS) (according to Yao et al. (2014))*

3.1.5. *Porphyrans*

Porphyrans are sulfated polysaccharides present in red algae of the genus *Porphyra*. Algae belonging to this genus are found to be richer in polysaccharides (76% of DM) than other species (Holdt and Kraan 2011). Among this polysaccharide fraction, porphyrans represent the major compound (Qiu et al. 2021). *Porphyra tenera* and *Porphyra yezoensis* species containing porphyrans are frequently used in human foods under the name "nori" and are used in many products, including maki. The basic structure of porphyrans is a disaccharide consisting of an L-galactose-6-sulfate and a D-galactose-6-methyl linkage (see Figure 3.27).

By repeating this basic unit, the final polysaccharide, or porphyran, is created. Porphyrans, like carrageenans, show a wide range of molecular weights. In the species *P. yezoensis*, the porphyran fraction is characterized by an average mass value of $1.8.10^5$ Da, with a maximum value of $2.4.10^5$ Da (Qiu et al. 2021). Porphyrans and their oligosaccharide derivatives are associated with numerous biological activities of therapeutic interest (see Figure 3.28).

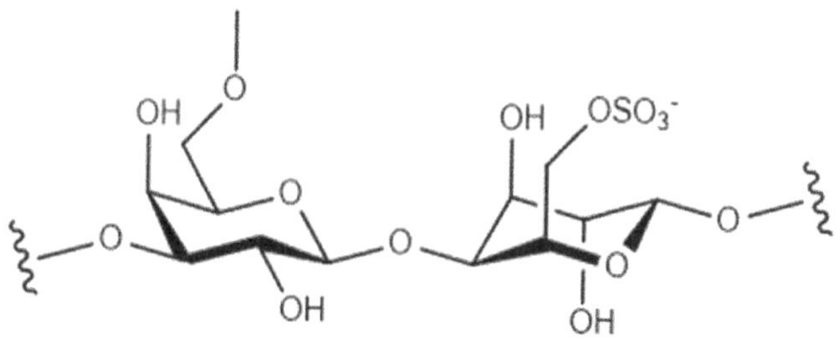

Figure 3.27. *Porphyran structure (according to Fleurence (2018))*

Oligosaccharides are obtained either chemically or enzymatically. By chemical means, porphyranes can be hydrolyzed in an acid medium at temperatures ranging from 60°C to 90°C. The oligosaccharide fractions obtained by this process are characterized by low molecular weights, ranging from 1,430 to 6,120 Da. The enzymatic pathway is based on the use of agarase or porphyranase. The latter is more specific to porphyran hydrolysis and is gradually replacing agarase.

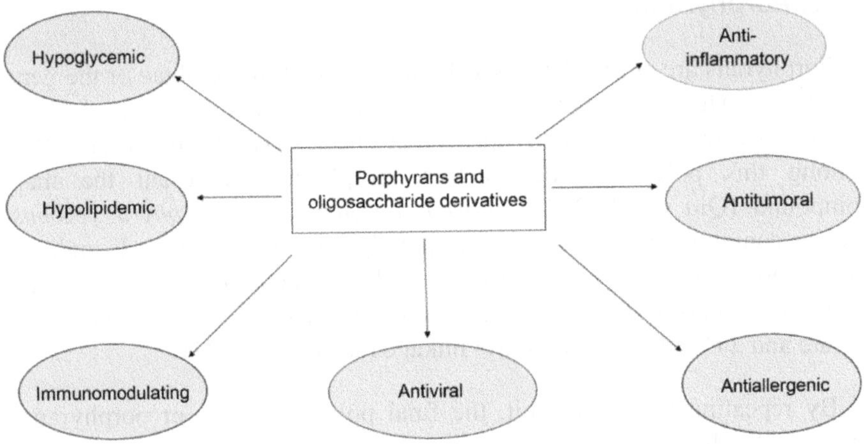

Figure 3.28. *Some examples of biological activities of human health interest associated with porphyrans and their oligosaccharide derivatives (according to Qiu et al. (2021))*

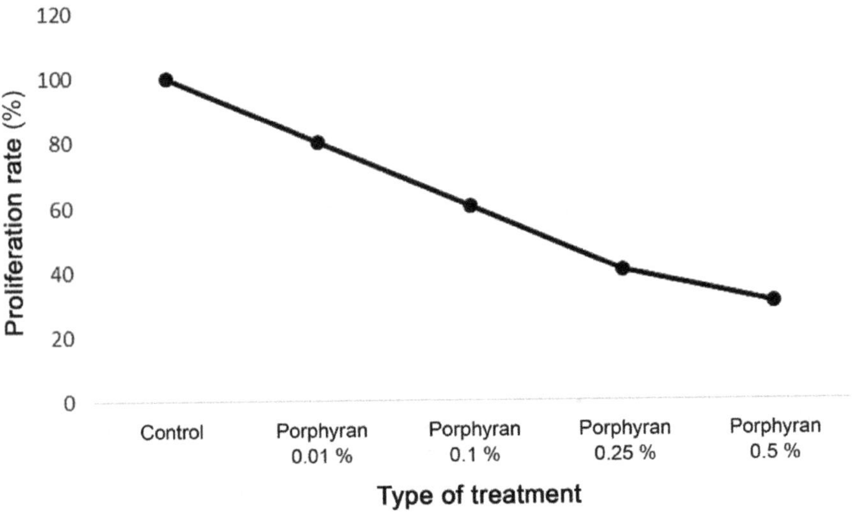

Figure 3.29. *In vitro effect of the addition of increasing doses of porphyrans on AGS tumor cell cultures (according to Kwon and Nam (2006))*

Porphyrans have been shown to induce apoptosis in the AGS tumor line of human gastric adenocarcinoma. These polysaccharides do not affect the growth of normal cells but induce the inhibition of malignant cell proliferation and cell death when supplied at increasing doses (see Figure 3.29) (Kwon and Nam 2006).

Porphyrans are also known for their anti-inflammatory and antiallergenic activities (Fleurence and Ar Gall 2016). These activities are mainly described from dried nori, which is a traditional food product in Asia. In addition to these activities, porphyrans are also described for their regulatory effects on glycemia and lipodemia. Their inclusion in a rat diet leads to a significant reduction in serum triglycerides (-29%), compared with a high-fat diet. Similarly, their dietary intake induces in rats a significant reduction in "bad cholesterol" (LDC) (-16%) and an increase in "good cholesterol" (HDC) (+30%) (Qian et al. 2014). In addition to porphyrans, their depolymerized derivatives also carry on therapeutic activity.

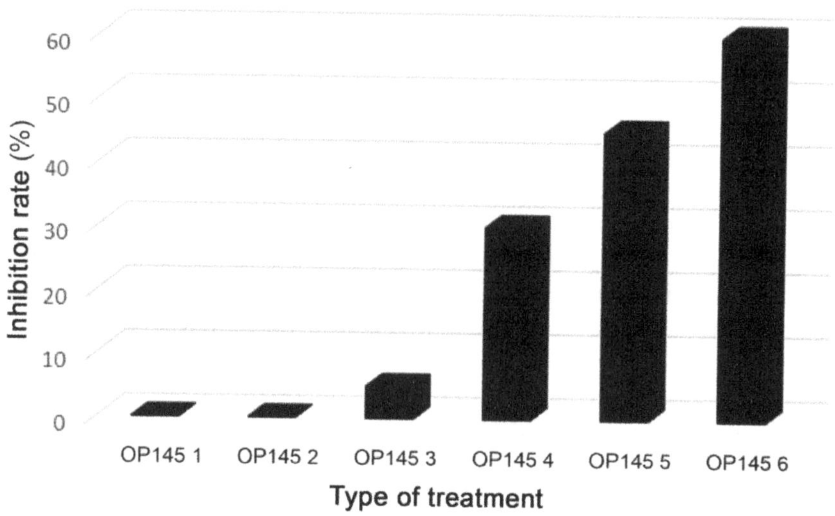

Figure 3.30. *Effect of different doses of OP145 oligosaccharides on inhibition of SARS-CoV-2 virus attachment to target cells (OP 145 1: 0.390 µg/mL; OP 145 2: 1.563 µg/mL; OP 145 3: 6.25 µg/mL; OP 145 4: 25 µg/mL; OP 145 5: 100 µg/mL; OP 145 6: 400 µg/mL) (according to Geng et al. (2023))*

Recent work has highlighted an antiviral activity of major interest for human health with an oligo-porphyran (Geng et al. 2023). This inhibits infection by the SARS-CoV-2 virus, responsible for Covid-19. This type of inhibitory activity has already been reported for other sulfated polysaccharides such as carrageenans (Frediansyah 2021).

The oligo-porphyran tested has a molecular mass of 2,100 Da and is derived from the alga *Pyropia yezoensis* (*Porphyra yezoensis*). It is obtained by acid hydrolysis of the native porphyridane, which has an initial mass value of 133,450 Da. Listed as OP145, this oligosaccharide shows significant anti-SARS-CoV-2 activity. Indeed, its IC_{50} is established at 37.52 µg/mL and its inhibiting activity increases with the dose administered (see Figure 3.30). Its mechanism of action also seems to have been elucidated. The oligosaccharide appears to have an affinity for the viral S protein and binds to it. This mechanism of neutralization of protein S by the oligo-porphyran thus limits viral attachment, and hence penetration of target cells.

In addition to antiviral activity, porphyran oligosaccharides also appear to have neuroprotective activity. This type of activity, highlighted in the murine model, relates to a progressing neurodegenerative pathology, namely Parkinson's disease (Liu et al. 2018). The oligosaccharides used in this study have molecular masses ranging from 409 to 517 Da. Their use limits the decline in dopamine production observed after the application of 1-methyl-4-phenyl-1,2, 3,6-tetrahydropyridine (MPTP), a toxic molecule stimulating Parkinsonian disorders (see Table 3.10). They also stop the loss of dopaminergic neurons, neural cells involved in dopamine synthesis. In the presence of MPTP, almost 50% of these cells are lost, whereas the loss is reduced to 20% after administration of oligo-porphyrans at a dose of 50 mg/kg body weight. MPTP also acts markedly on the dopamine transporter and receptors, while treatment with porphyran oligosaccharides remedies these MPTP effects. At a concentration of 50 mg/kg body mass, these oligosaccharides reduce degenerative effects in a manner comparable to Madopar, which is a clinical antiparkinsonian treatment (Liu et al. 2018).

Type of treatment	Indicator	MPTP	Oligoporphyrans (25 mg/kg)	Oligoporphyrans (50 mg/kg)
Amount of dopamine in striatum (μg/g)	4	2.2	3	3.3

Table 3.10. *Effect of the action of oligosaccharides porphyrans on the amount of dopamine present in the striatum cerebral gland in remediation of the submission of mice to an MPTP Parkinsonian inducer (according to Liu et al. (2018))*

Porphyrans have biological activities that could be exploited in pharmaceutical research and development. However, the molecular weight of these polysaccharides, as with carrageenans, can pose a problem for the formulation and administration of these active ingredients. The current focus on the search for activities carried by depolymerized forms of porphyrans therefore appears to be a fully justified new research strategy.

3.1.6. *Ulvans*

Ulvans are the main constituent of the cell wall of green algae. They make up 8–29% of the dry matter of algae belonging to this family (Véra et al. 2011). As with other algal polysaccharides, ulvans have a complex structure. This is based on the linking of a basic disaccharide repeated n times. This basic structure is composed of the association of an L-rhamnose-3 sulfate with a guluronic acid (see Figure 3.31). This unit is called "ulvabiouronic acid A". Its association with other ulvabiouronic acids leads to the formation of highly heterogeneous ulvans (Fleurence 2018).

Ulvans, like other algal polysaccharides, can also be depolymerized, mainly by chemical or enzymatic hydrolysis or physical treatment (ultrasonication). Sonication treatment yields oligo-ulvans with molecular weights, some of them between 50 and 60 kDa (Véra et al. 2011).

Like many sulfated algal polysaccharides, ulvans carry biological activities of interest, such as anticancer, immunomodulatory or hypolipidemic activities (Li et al. 2023).

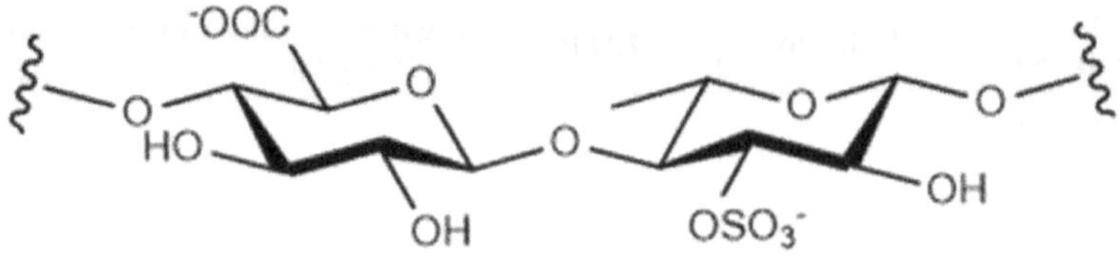

Figure 3.31. *Basic structure of ulvabiouronic acid (according to Fleurence (2018))*

The anticancer activity of ulvans seems to be based on three mechanisms: triggering cell apoptosis, reducing malignant cell growth and activating the innate immune response. This type of antitumor action has been demonstrated for ulvans extracted from *Ulva fasciata*, commonly used in human food as "sea lettuce" (Shao et al. 2014). The ulvan fractions, obtained by hot water extraction and separated by ultrafiltration, are listed as "UF1", "UF2" and "UF3". They are distinguished by very different molecular weights, ranging from 1,900 Da for UF1, to 54,700 Da for UF2 and 262,700 Da for UF3. They are also characterized by very different sulfate and uronic acid contents (see Figure 3.32).

These fractions were tested on malignant DLD human colorectal cancer cells. The highest antitumor activity was reported for the UF3 fraction, which differs from the other two in having a lower degree of sulfation and a higher uronic acid content than the other fractions (see Figure 3.32).

In addition to their antitumor activity, ulvans are also described in the literature for their antiviral properties and antiulcer activity (Holdt and Kraan 2011). These ulvans are also derived from green algae frequently consumed as sea vegetables, namely *Ulva lactuca*, *Ulva rigida* (sea lettuce) and *Monostroma* sp. (aonori).

Paradoxically, there is little information on the therapeutic activities associated with depolymerized forms of ulvans or oligoulvans. Most scientific articles focus on the eliciting activities of oligoulvans on plants' natural infection mechanisms, whether viral, bacterial or fungal in origin. This type of observation suggests that the main focus of oligoulvan development currently remains plant health.

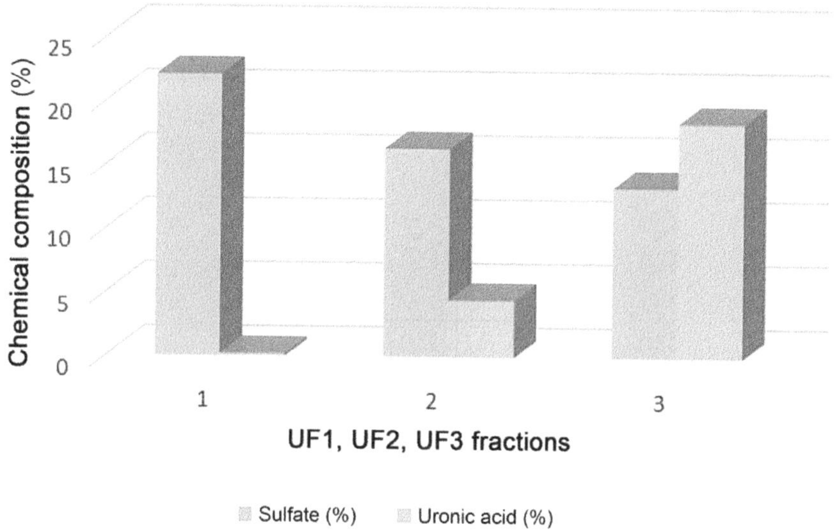

Figure 3.32. *Sulfate and uronic acid composition of ulvan fractions UF1 (1), UF2 (2) and UF3 (3) extracted with hot water from* Ulva fasciata *(according to Shao et al. (2014)). For a color version of this figure, see www.iste.co.uk/fleurence/health.zip*

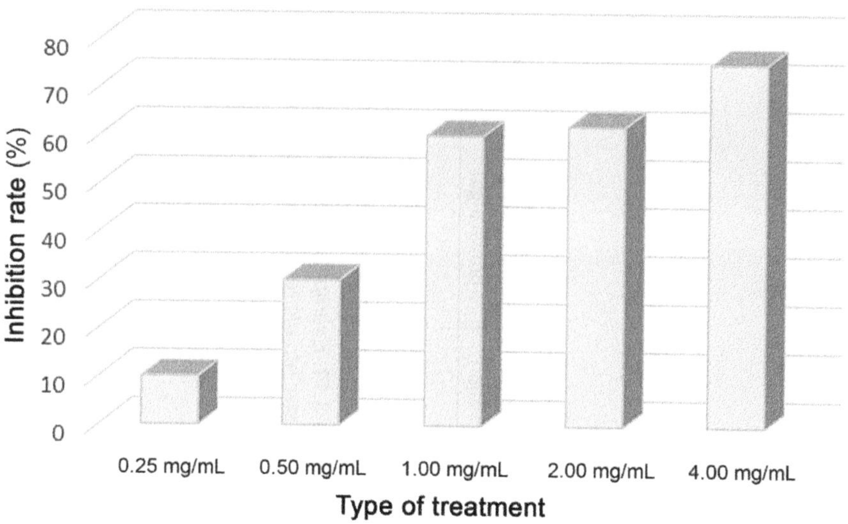

Figure 3.33. *Effect of increasing doses of fraction 3 (UF3) of ulvans on the growth inhibition rate of the DLD colorectal cancer cell line (according to Shao et al. (2014))*

Species	***Laminaria japonica*** (brown alga)	***Undaria pinnatifida*** (brown alga)	***Eisenia bicyclis*** (brown alga)	***Porphyra tenera*** (red alga)	***Palmaria palmata*** (red alga)	***Gelidium amansii*** (red alga)	***Ulva lactuca*** (green alga)
Protein content (% of DM)	13.0–15.4	11–24	18.5	33–47	8.1–35.6	18.1	8.7–33.7

Table 3.11. *Examples of protein content and their variations according to botanical families (brown algae; red algae, green algae) and species (according to Ochia et al. 1986; Fleurence 2024)*

3.2. Proteins and peptides

The protein content of algae can vary according to major botanical classes (Chlorophyceae, Rhodophyceae, Phaeophyceae) and species (see Table 3.11) (Ochiai et al. 1985; Fleurence 1999; Fleurence 2024). The protein content of algae is also subject to strong seasonal fluctuations. However, these variations are not always studied, and this type of data may be incomplete for certain species (see Table 3.11).

The most protein-rich algae are mainly Rhodophyceae. This is due to the presence of a protein supernumerary pigment characteristic of this class. This pigment, known as R-phycoerythrin (RPE), is a high molecular weight protein (240 kDa). It consists of a stack of several subunits, called α (20 kDa), β (21 kDa) and γ (30 kDa), and is a member of the phycobiliprotein family. It is an oligomeric protein whose final structure is a trimer of the basic αβ dimer on which a γ subunit associates, i.e. a final structure $[(\alpha\beta)_6\ \gamma]$. R-PE alone can account for 12% of the total protein fraction at certain seasons (Galland-Irmouli et al. 2000).

a) b)

Figure 3.34. *R-phycoerythrin (R-PE) (b) extracted from the red alga* Palmaria palmata *(a) (photo credit © Fleurence J. 2024). For a color version of this figure, see www.iste.co.uk/fleurence/health.zip*

This purple-pink protein (see Figure 3.34) is characterized by three peak maxima of light absorption, located at 499 nm and 545 nm respectively (Galland-Irmouli et al. 2000). The light absorbed by the RPE is restored in the form of fluorescence at the single wavelength of 578 nm.

This fluorescence emission sometimes induces biological activity, as will be detailed later.

Both the native protein and the subunits may carry activities of therapeutic interest (antiparasitic, antitumor). The antitumor activity of R-PE is often associated with the protein pigment's constituent subunits (Bei et al. 2002; Ulagesan et al. 2021).

In particular, the α-subunit has been described as an anticancer agent against the HepG2 liver tumor line at concentrations between 5 and 30 μg/mL. At these concentrations, this subunit proves cytotoxic against malignant cells (see Figure 3.35) (Ulagesan et al. 2013).

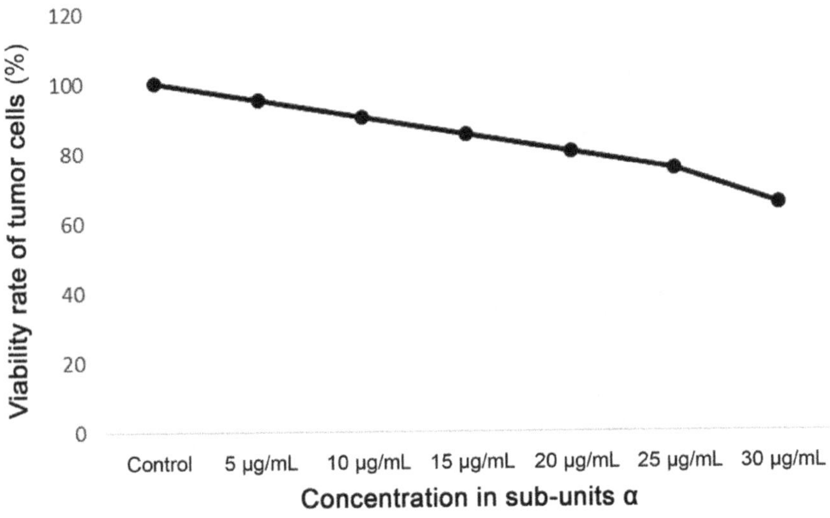

Figure 3.35. *Impact of different concentrations of α-subunits of R-PE obtained from the red alga* Pyropia yezoensis (Porphyra yezoensis) *on the viability of liver tumor cells of the HepG2 line (according to Ulagesan et al. (2021))*

The impact of R-PE subunits in the treatment of cancer cells was also the subject of an original study published in the early 2000s (Bei et al. 2002).

The three subunits α, β, γ, when irradiated with an argon laser at a wavelength of 496 nm and an energy of 28 J/cm^2, triggered an apoptosis process in tumor cells (Sarcoma 180 line, hepatic carcinoma SMC 7721 line) having internalized said subunits. Moreover, the phototoxicity recorded for tumor cells did not appear to affect normal cells.

The antitumor activity of native R-PE is also well described. The latter has been reported in vitro and in vivo in two different models. The first is a study carried out on the Hela cell line (breast cancer) for the in vitro aspect and work carried out on Sarcoma 180 implanted in the murine model for the in vivo side (Pan et al. 2013). The R-PE tested in this study is derived from the species *Porphyra haitanensis*.

The antitumor activity of R-PE in Hela cells is evaluated without or with laser irradiation (ʎ excitation: 632.8; energy 12.5 J/cm^2) of tumor cell cultures that have internalized the pigment.

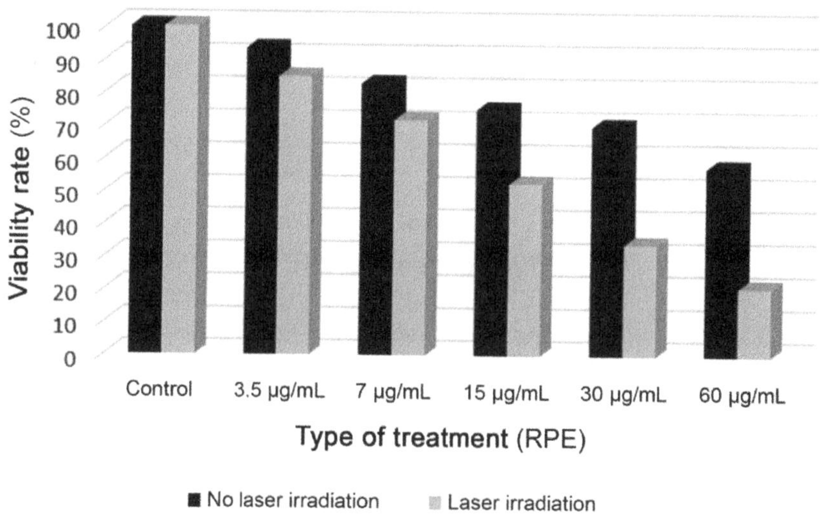

Figure 3.36. *Effect of different R-PE concentrations on the viability rate of Hela cells subjected or not to laser irradiation (irradiation wavelength 632.8 nm, energy submitted 12.5 J/cm^2) (according to Pan et al. (2013))*

The R-PE/irradiation combination can inhibit Hela cell viability by up to 80% when the dose of R-PE administered is 60 µg/mL (see Figure 3.36).

As before, irradiation of integrated R-PE by tumor cells induced apoptosis. This remarkable result highlights the value of photodynamic therapy and the use of R-PE or its subunits as anticancer agents. This finding seems to encourage the development of cancer therapies other than chemotherapy.

The in vivo effect of R-PE on the Sarcoma 180 solid tumor is dose-dependent (see Figure 3.37). It is optimal for an injected R-PE concentration of 300 mg/kg. At this dose, the tumor growth inhibition rate is 41.3%.

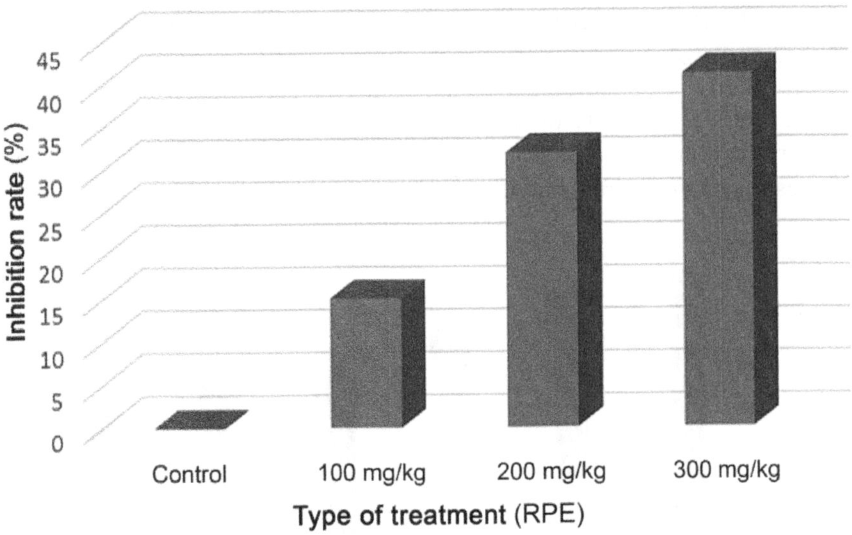

Figure 3.37. *Effect of increasing concentrations of R-PE on the inhibition of Sarcoma 180 tumor growth implanted in a murine model (according to Pan et al. (2013))*

R-PE therefore appears to be a molecule capable of inhibiting the growth of a solid tumor such as Sarcoma 180, without the need for photoinduction by laser irradiation.

The phycobiliprotein family (phycoerythrin, phycocyanin, allophycocyanin) is associated with numerous biological activities. These include antiviral, antitumoral, anti-inflammatory, but equally antioxidant, neuroprotective and antidiabetic activities (Thiviya et al. 2022). The phycobiliproteins concerned by these properties are obtained from red algae such as *C. crispus*, *Chondrus armata*, *Gracilaria* sp., *Gelidium* sp. and

Palmaria palmata. These species are used as sea vegetables for human consumption.

Apart from pigment proteins, such as phycobiliproteins, algae possess other proteins with interesting therapeutic properties. These include proteins linked to various oligosaccharidic chains or glycans. These glycoproteins may be located in the cell wall, on the cell surface or excreted into the environment. They may therefore play a role in cell interactions and recognition. The composition of these glycoproteins varies widely. The protein fraction may be in the minority compared with the glycosylated fraction, as is the case for the glycoproteins of algae of the genus *Ulva* (see Figure 3.38) (Echave et al. 2022).

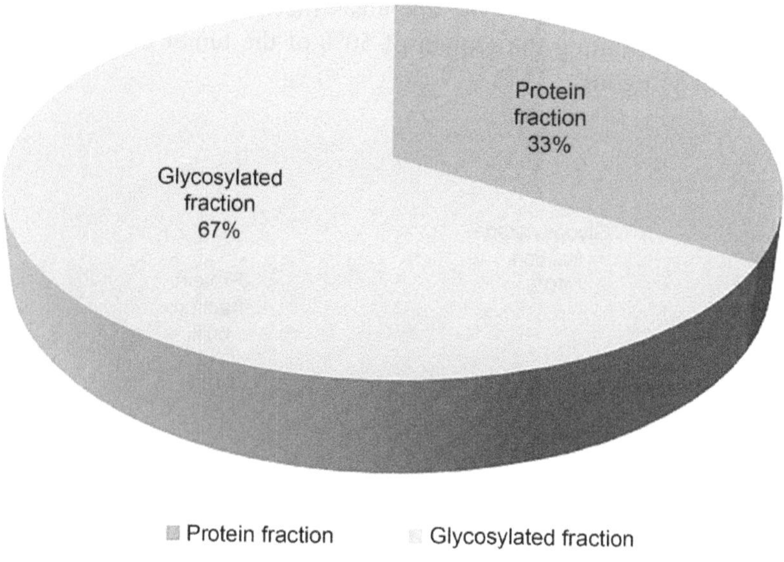

Figure 3.38. *Distribution of the glycosylated fraction and the protein fraction within the glycoproteins of algae of the genus* Ulva *(according to Echave et al. (2022))*

In contrast, green algae of the genus *Codium*, and more specifically the species *Codium decorticatum*, have glycoproteins in which the protein fraction is predominantly represented in relation to the glycosylated fraction (see Figure 3.39).

Algal glycoproteins therefore have different compositions and structures, which give rise to a variety of biological activities. A well-known group of these glycoproteins are lectins. Lectins are glycoproteins involved in cell recognition mechanisms between organisms. They are often described for their affinity to bind preferentially to mannose-rich oligosaccharide motifs. Lectins and other algal glycoproteins are listed in the literature for their numerous properties of therapeutic interest, which will be summarized later (see Table 3.12) (Echave et al. 2022).

The glycoprotein fraction of the green alga *C. decorticatum* is characterized by the presence of a glycoprotein with a molecular mass of 48 kDa and antitumor activity towards the lung cancer cell line MDA-MB-231 (Thangam et al. 2014). This glycoprotein, which has a cytotoxic effect on tumor cells, notably triggers the apoptosis mechanism in tumor cells. The concentration inhibiting the growth of 50% of the tumor cell population is estimated at 55 µg/mL.

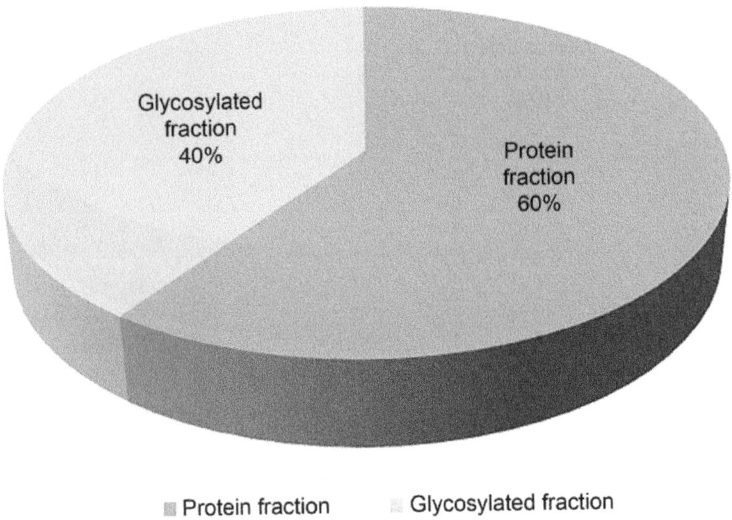

Figure 3.39. *Distribution of glycosylated fraction and protein fraction within glycoproteins in* Codium decorticatum *species (according to Echave et al. (2022))*

A comparable result was obtained with the lectin derived from the red alga *Eucheuma serra* on osteosarcoma cell lines OST and LM8. This lectin, with its agglutinating properties, is known as ESA (*Eucheuma serra* agglutinin). ESA lectin is toxic to osteosarcoma tumor cells. This antitumor activity is reflected in a significant decrease in cell viability in the two lines, which is greatest at an ESA concentration of 50 µg/mL (see Figure 3.40).

In addition to their antitumor activity, lectins are also involved in inhibiting the attachment of viral particles to the membrane of cells targeted by the virus. Lectin, or griffithsin (GRFT), obtained from the red alga *Griffithsia* sp., inhibits infection by HIV-1, HSV and HPV viruses in vivo by this mechanism. GRFT is a small glycoprotein with a molecular mass of 12.7 kDa and 121 amino acids. In the case of HIV, GRFT binds to the viral envelope at the level of the oligosaccharides of the gp 120 protein. It blocks the latter and prevents it from binding to the CD4 receptor on the T cell membrane, thus blocking the virus's ability to penetrate the cell (see Figure 3.41) (Mori et al. 2005).

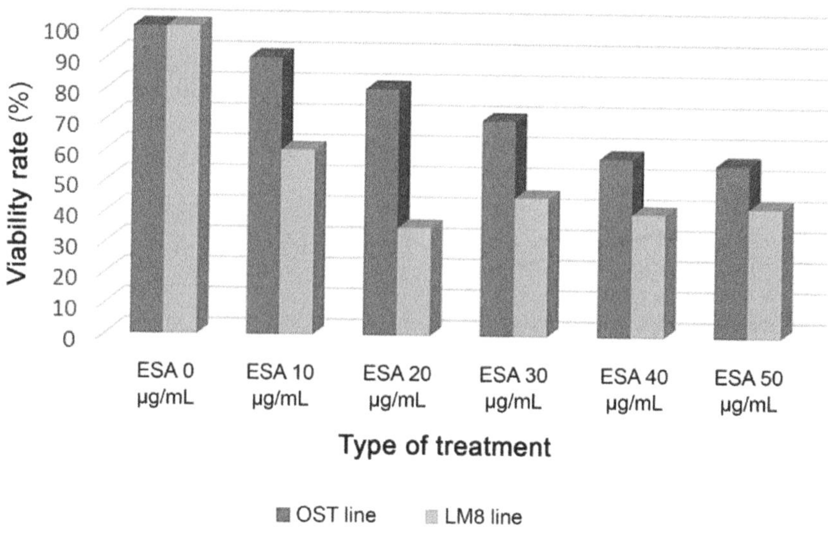

Figure 3.40. *Effect of adding increasing concentrations of the lectin ESA on cell viability in OST and LM8 tumor lines (according to Hayashi et al. (2012))*

Moreover, the same lectin is described in the literature as having a binding affinity with the S glycoprotein of viruses in the coronavirus class (Echave et al. 2022). This type of property has been known since the early 2000s, when research focused on the investigation of molecules likely to counteract the severe acute respiratory syndrome (SARS) epidemic that raged in China and Canada between 2000 and 2004.

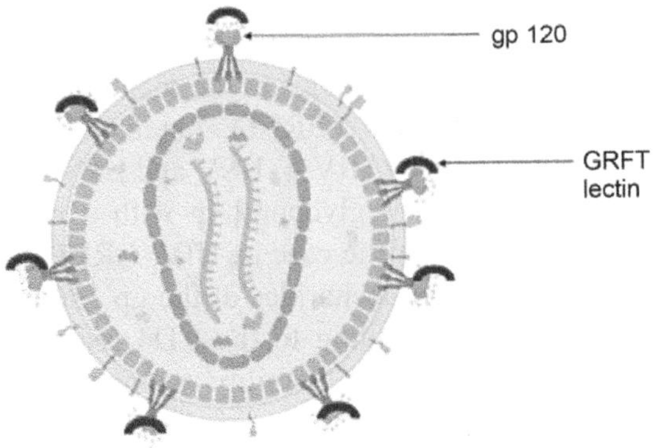

Figure 3.41. *Inhibition mechanism of the HIV viral glycoprotein gp 120 by binding of the algal lectin GRFT from the red alga* Griffithsia *sp. (according to Mori et al. (2005*[3]*) (modified by Fleurence (2024))*

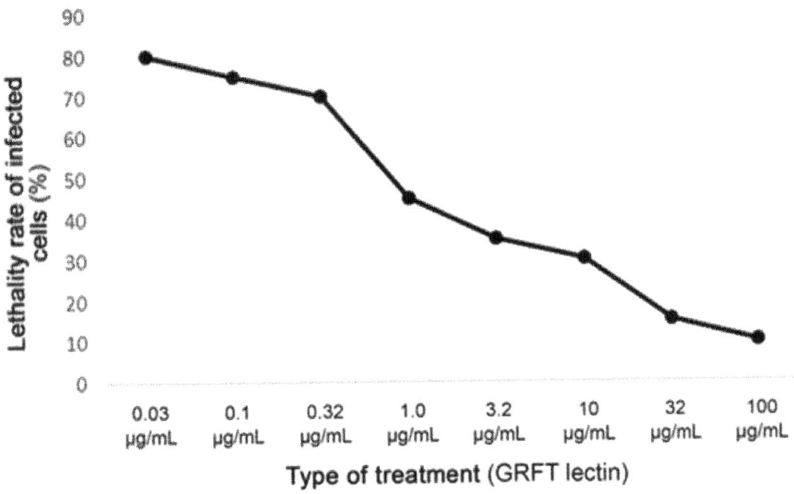

Figure 3.42. *Effect of increasing concentrations of the algal lectin GRFT on the lethality of Vero 76 cells infected with SARS-CoV Urbani virus (according to O'Keefe et al. (2010))*

3 See: https://commons.wikimedia.org/wiki/File:HI-virion-structure_en.svg.

In particular, GRFT lectin has been shown to be effective against SARS-CoV-1 infection and the SARS-CoV Urbani strain (O'Keffe et al. 2010). It limits the cytopathic effect of the virus on infected Vero 76 cells (see Figure 3.42). It significantly reduces the number of cells killed by SARS-CoV Urbani. The effect on cell lethality was dependent on the concentration of GRFT administered to the cell culture (see Figure 3.42). The median effective concentration (EC_{50}) to induce 50% inhibition of viral infection by SARS-CoV Urbani was set at 0.61 μg/mL lectin.

GRFT lectin has been shown to have an affinity for mannose-containing oligosaccharides carried by HIV gp 120 and coronavirus Spike (S) protein. This affinity is significantly higher for HIV gp 120, with a dissociation constant (Kd) of 8.2 nM, compared with 24.9 nM for SARS-CoV protein S (O'Keefe et al. 2010). This difference in affinity translates into a very different stoichiometric ratio between GRFT and the two types of viral proteins (see Figure 3.43). The stoichiometric ratio between GRFT and gp 120 is 10, compared with 3 for S protein. This suggests that the gp 120 protein contains more mannose-rich oligosaccharide sites than the S protein. This type of site is indeed at the root of the recognition mechanism, and therefore of the binding of GRFT with viral envelope proteins.

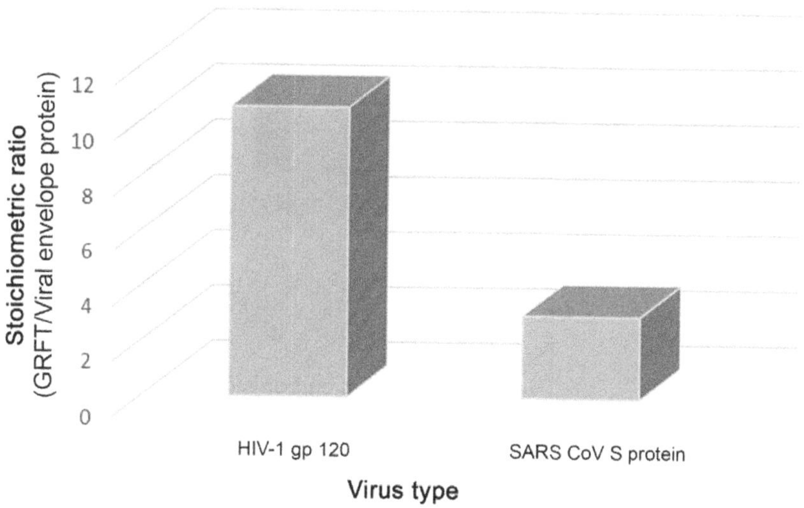

Figure 3.43. *Stoichiometric binding ratio between GRFT lectin and the glycosylated sites of the gp 120 proteins of the HIV virus and the S protein of SARS CoV (according to O'Keeffe et al. (2010))*

Lectins, like GRFT, have several mechanisms of action that may explain the antiviral effects observed. Among these, inhibition of virus penetration through obstruction of the glycoproteins that attach the virus to the cell membrane remains an essential mechanism. Despite this, GRFT, like other lectins, can have a virucidal effect (Echave et al. 2022).

More generally speaking, algal glycoproteins and lectins are associated with numerous activities of major therapeutic interest (see Table 3.12). In addition to their therapeutic properties, lectins have also been described for their erythrocyte agglutination properties. This is the case for the lectin derived from the green alga *Ulva lactuca*, which shows a very high affinity for group O human erythrocytes and a lower hemagglutination activity for groups B, A and AB (Gilboa-Garber et al. 1988). This property is based on the recognition by lectins of certain saccharide motifs carried by erythrocyte cell membranes. In the case of ulva, the lectin specifically recognizes L-Fucose and D-Galactose. The addition of these two carbohydrates to the medium containing erythrocytes and algal lectin inhibits the hemagglutination process (see Figure 3.43).

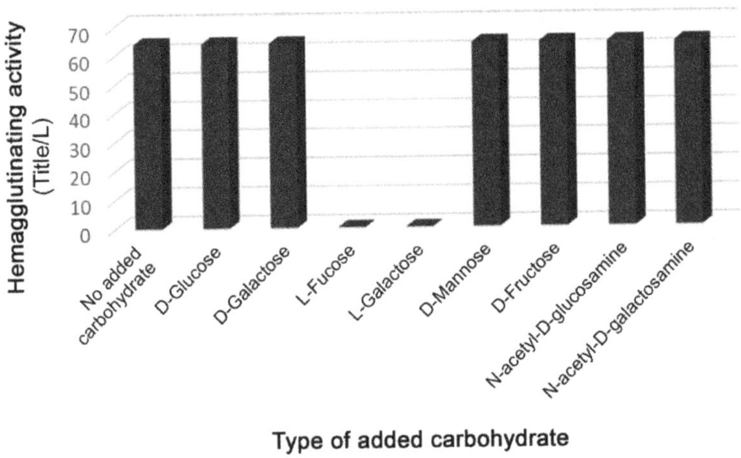

Figure 3.44. *Hemagglutinating activity of* Ulva lactuca *lectin (1 mg/mL) in the presence of different types of added carbohydrates (0.2%) (according to Gilboa-Garber et al. (1988))*

This property of recognizing saccharide motifs on the surface of cellular membranes makes algal lectins interesting reagents for studying membrane biochemical composition. Like lectins from terrestrial plants or antibodies, they are potential tools for cell typing, which remains a major focus of biomedical research.

Algal proteins whether phycobiliproteins or lectins, present biological activities that could have applications in the field of human health. This also applies to their degraded or derived forms, peptides.

Peptides obtained from enzymatic hydrolysis of algal proteins demonstrate numerous biological activities likely to be of interest to medical research. These often include hypotensive, antidiabetic and, more secondarily, anti-inflammatory, antitumoral, anticoagulant, antioxidant or even antimicrobial activities (see Table 3.13) (Echave et al. 2022; Leong and Chang 2024).

Algal species	Bioactive protein	Activity	References
Caulerpa cupressoides (green alga)	Lectin	Anti-inflammatory Pain relief	Thiviya et al. (2020) Echave et al. (2022)
Codium fragile (green alga)	Lectin	Lipogenic	Smit (2004)
Boodlea coacta (green alga)	Lectin	Antiviral (hepatitis C, SARS CoV)	Thiviya et al. (2020) Echave et al. (2022)
Ulva fasciata (green alga)	Lectin	Hemagglutinating	Kumar and Barros (2020)
Sargassum fusiform (brown alga)	Lectin	Antioxidant	Echave et al. (2022)
Undaria pinnatifida (brown alga)	Glycoprotein	Antidiabetic	Echave et al. (2022)
Laminaria japonica (brown alga)	Glycoprotein	Antioxidant	Echave et al. (2022)
Eucheuma denticulatum (red alga)	Lectins (3 isoforms)	Antimicrobial (*Vibrio alginolyticus*)	Hung et al. (2014)
Solieria filiformis (red alga)	Lectins (SfL-1, SfL-2)	Antitumor Antidepressant	Maliki et al. (2022)
Bryothamnion triquetrum (red alga)	Lectin (BTL)	Cancer biomarkers	Maliki et al. (2022)

Table 3.12. *Some examples of activities of therapeutic interest associated with algal lectins*

Algal species	Protease used	Peptide sequence	Biological activity	References
***Porphyra* spp.** (red alga)	Pepsin (pH 2, 45°C)	GGSK, ELS	Antidiabetic	Echave et al. (2022)
Porphyra dioica (red alga)	Alcalase + Flavourzyme	YLVA	Antidiabetic	Leong and Chang (2024)
Palmaria palmata (red alga)	Chymotrypsin (pH 8, 30°C)	33 bioactive peptides < 10 kDa	Hypotensive Antidiabetic	Echave et al. (2022) Leong and Chang (2024)
Gracilariopsis chorda (red alga)	Thermolysine	IDHY, LRY, LVVER	Hypotensive	Leong and Chang (2024)
***Ulva* spp.** (green alga)	Protease alkaline + Flavourzyme	SLAVSVH	Antidiabetic	Leong and Chang (2024)
Ulva rigida (green alga)	Pepsin (pH 2, 37°C)	IP, ALF	Hypotensive	Echave et al. (2022)
***Ulva intestinalis* (*Enteromorpha intestinalis*)** (green alga)	Trypsin (pH 8, 37°C)	FGMPLDR, MELVR	Hypotensive	Sun et al. (2009) Echave et al. (2022) Leong and Chang (2024)
Laminaria digitata (brown alga)	Papain + Alkalase + Pepsin	FY, AKYSY	Hypotensive	Leong and Chang (2024)
Undaria pinnatifida (brown alga)	Bromelain	KNFL	Hypotensive	Echave et al. (2022)

Table 3.13. *Some examples of methods for obtaining algal peptides by enzymatic hydrolysis and biological properties of therapeutic interest*

The production of these peptides is mainly achieved through the use of various proteases (see Table 3.13). The conditions under which hydrolysis takes place, such as pH and temperature, vary from enzyme to enzyme (see Table 3.13). However, most durations remain relatively long (5–24 h).

The mechanism of action involved in the hypotensive properties of algal peptides is based on their inhibition of the angiotensin-converting enzyme (ACE). This inhibitory effect has been demonstrated using peptides from various species, including the green alga *Ulva intestinalis* (Sun et al. 2019).

In this species, enzymatic hydrolysis of the protein fraction via trypsin gives the highest angiotensin-conversion inhibitory activity. The peptides from this hydrolysis are fractionated by ultrafiltration, and each fraction is characterized by its molecular weight. The fraction containing peptides with a molecular mass of less than 3 kDa is the most effective in inhibiting ACE activity (see Figure 3.45). The inhibition rate observed with this fraction was 53.01%. The minimum concentration to achieve 50% inhibition of ACE activity (IC_{50}) is 1.14 mg/mL (Sun et al. 2019) (see Figure 3.46).

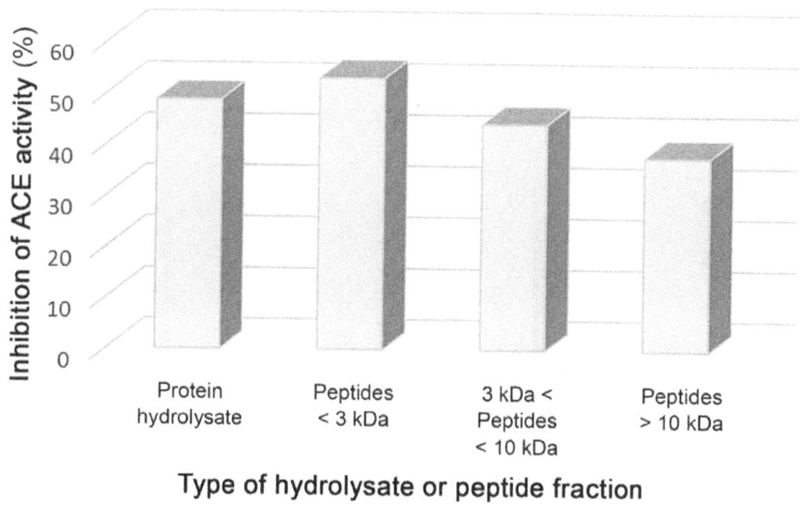

Figure 3.45. *Inhibitory effect on ACE activity of various peptide fractions derived from enzymatic hydrolysis of the green alga* Ulva intestinalis *(according to Sun et al. (2019))*

Purification by gel filtration chromatography of the peptide fraction with a molecular mass of less than 3 kDa revealed two peptides of sequence FGMPLDR and MELVLR with molecular masses of 834.41 and 799.43 Da respectively. These two purified peptides showed an IC_{50} of 0.183 mg/mL and 0.179 mg/mL respectively, compared with 1.14 mg/mL with the parent fraction (< 3 kDa) (see Figure 3.46).

Diabetes, and type II diabetes in particular, poses a real public health problem worldwide. It affects almost 9% of the adult population (Balansa 2023). A recent study shows that the administration of a protein hydrolysate in streptozotocin-induced diabetes mice significantly improved glycemic and lipid profiles (Mc Laughlin et al. 2020). Adding protein hydrolysate to the diet of diabetic animals twice a day reduced non-fasting blood glucose levels by almost 29%. This reduction is similar to that observed when taking metformin, a first-line treatment for type II diabetes. *P. palmata* peptide solution also had a positive effect on plasma insulin levels. Plasma insulin levels were increased by around 67% compared with those in untreated diabetic mice. A comparable result was observed when metformin, a drug used to treat type II diabetes, was administered.

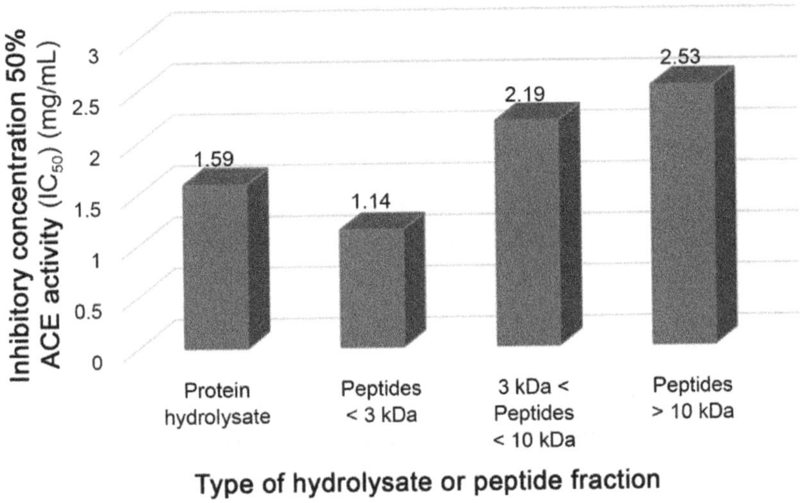

Figure 3.46. *Determination of the inhibitory concentration inhibiting 50% of ACE activity by peptide fractions derived from the enzymatic hydrolysis of the protein fraction of the green alga* Ulva intestinalis *(according to Sun et al. (2019))*

Algal species	**Total lipids** (% of DM)	**Neutral lipids** (% total fatty acid fraction)	**Glycolipids** (% total fatty acid fraction)	**Phospholipids** (% total fatty acid fraction)	**References**
Ulva rotundata (green alga)	1.95	46	37.9	16.1	Fleurence et al. (1994)
Ulva intestinalis (green alga)	1.85	30.3	55	14.7	Fleurence et al. (1994)
Laminaria digitata (brown alga)	1.47	59.2	29.3	11.5	Fleurence et al. (1994)
Saccharina latissima (brown alga)	3.24	66.7	24.4	10.9	Fleurence et al. (1994)
Fucus vesiculosus (brown alga)	2.36	67.8	26.2	6	Fleurence et al. (1994)
Undaria pinnatifida (brown alga)	6.48	15.1	51	33.9	Fleurence et al. (1994)

Algal species	Total lipids (% of DM)	Neutral lipids (% total fatty acid fraction)	Glycolipids (% total fatty acid fraction)	Phospholipids (% total fatty acid fraction)	References
Porphyra umbilicalis (red alga)	3.37	18.9	55.9	25.2	Fleurence et al. (1994)
Chondrus crispus (red alga)	0.61	19.8	56.5	23.7	Fleurence et al. (1994)
Palmaria palmata (red alga)	1.57	23	61.6	15.4	Fleurence et al. (1994)
Grateloupia turuturu (red alga)	3.3–4.1	20.1–41.8	42.3–46.8	11.2–33.4	Kendel et al. (2013)

Table 3.14. *Examples of the lipid composition of some edible algae from the English Channel and the Atlantic Ocean*

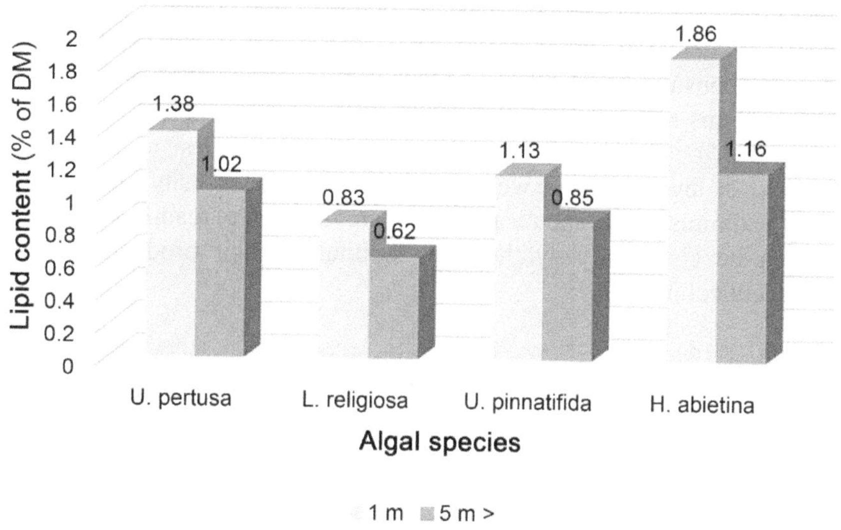

Figure 3.47. *Examples of the influence of depth (1 m or 5 m >) on the lipid content of several algal species* (Ulva pertusa, Laminaria religiosa, Undaria pinnatifida, Heterochordaria abietina) *(according to Ito and Tsuchya (1977))*

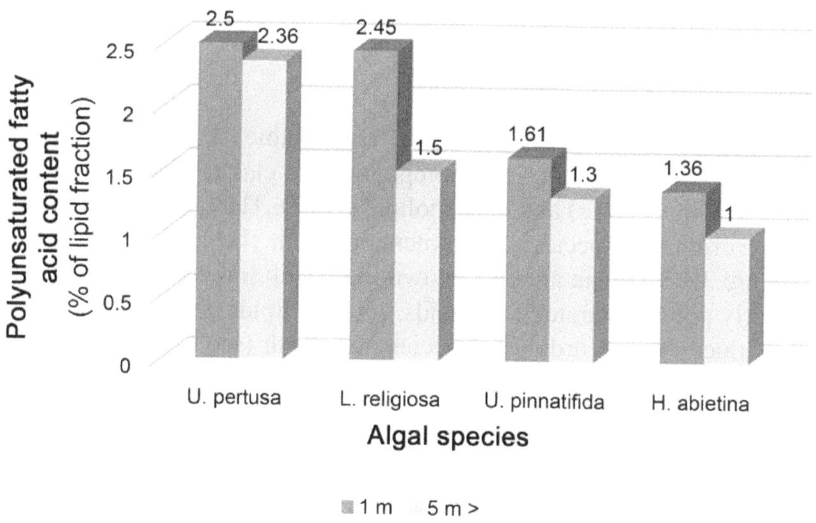

Figure 3.48. *Examples of the influence of depth (1 m or 5 m >) on the polyunsaturated fatty acid content of a number of algal species* (Ulva pertusa, Laminaria religiosa, Undaria pinnatifida, Heterochordaria abietina) *(according to Ito and Tsuchya (1977))*

Peptides are associated with numerous therapeutic activities of undeniable interest in human health. They are often characterized from algae frequently consumed in human diets. This finding rekindles the interest in consuming this marine plant resource for disease prevention (Fleurence and Levine 2016). Compared to polysaccharides, peptides also have the advantage of low molecular weight and low or zero viscosity, making them easier to administer therapeutically. They are, by nature, readily available, as they can be chemically synthesized, facilitating their production by the pharmaceutical industry.

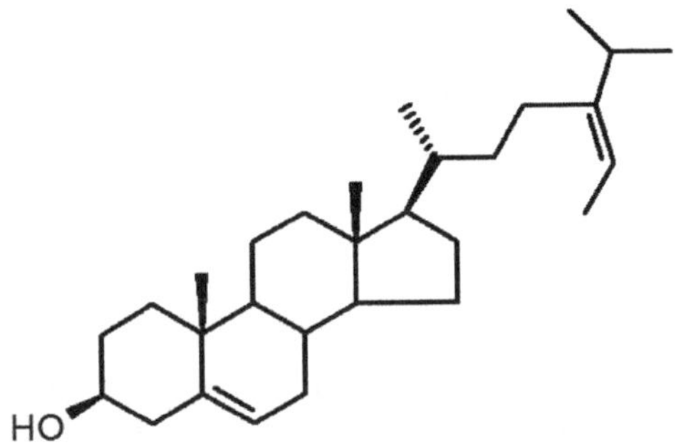

Figure 3.49. *Structure of the fucosterol molecule (according to Fleurence (2018))*

3.3. Lipids, fatty acids and sterols

Algae have a low lipid content (see Table 3.14). According to conventional nomenclature, lipid compounds are classified as neutral lipids (NLs), glycolipids (GLs) and phospholipids (PLs). The lipid content of algae varies according to species and immersion depth (Ito and Tsuchiya 1977) (see Figure 3.47). Algae are also known to contain low levels of fatty acids, particularly polyunsaturated fatty acids. As with lipids, the fatty acid content of algae fluctuates according to species and depth (see Figure 3.48). Other lipid-like substances are also present in algae. These include sterols, hydrophobic, unsaponifiable molecules belonging to the steroid class. Some of them, such as fucosterol, are associated with therapeutic properties that will be developed later (see section 3.3) (see Figure 3.49).

Numerous biological properties of therapeutic interest have been reported for lipids in the GL class. Antitumoral, anti-inflammatory and antimicrobial activities are particularly described (Lopes et al. 2021). GLs are complex and not always easy to analyze. However, several groups have been identified: galactolipids, glyceroglycolipids, sphingoglycolipids and sulfolipids. Galactolipids, which are widely represented in plants, are characterized by the presence of galactose linked to fatty acid chains. (see Figure 3.50). Depending on the number of galactose molecules involved in the association, a distinction is made between monogalactosyldiacylglycerols (MGDGs) (see Figure 3.50) and digalactosyldiacylglycerols (DGDGs).

Sulfoquinovosyldiacylglycerol (SQDG), although called a sulfolipid due to the presence of the sulfate group on the hexose (see Figure 3.51), also belongs to the GL class. Like other GLs (MGDG, DGDG), it carries biological properties of interest, such as antitumoral activities (see Table 3.15). This type of property is also found in PLs, which are polar lipids like GLs (see Table 3.15). PLs are polar lipids possessing a phosphoric acid group that gives them a hydrophilic property. PLs include phosphoglycerides and sphingolipids. Among the phosphoglycerides is phosphatidylcholine, a lipid first described in animals, and also present in algae (see Figure 3.52).

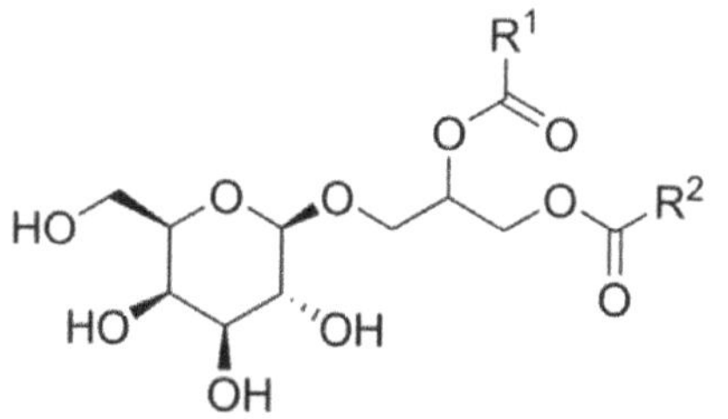

Figure 3.50. *Structure of monogalactosyldiacylglycerol (MGDG) (from https://commons.wikimedia.org/w/index.php?curid=8910083)*

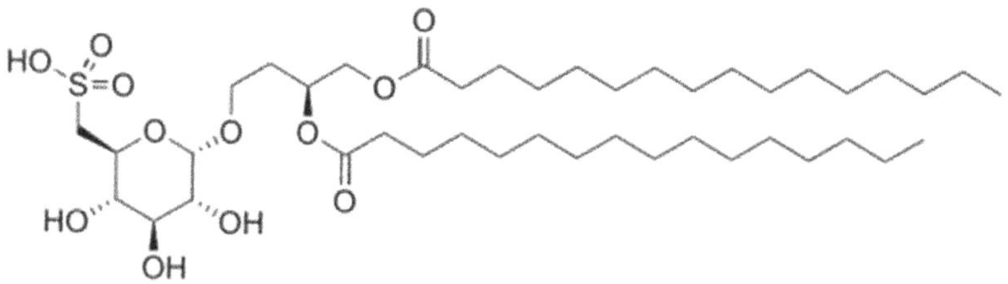

Figure 3.51. *Structure of sulfoquinovosyldiacylglycerol (SQDG[4])*

4 See: https://commons.wikimedia.org/w/index.php?curid=15256521.

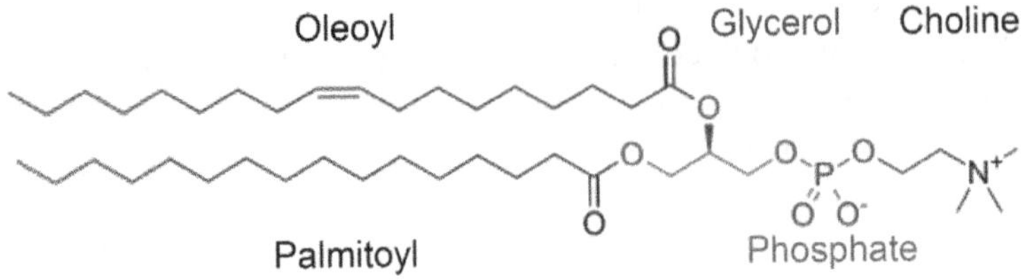

Figure 3.52. *Example of the structure of a phospholipid: phosphatidylcholine*[5]. *For a color version of this figure, see www.iste.co.uk/fleurence/health.zip*

GLs (MGDG, DGDG, SQDG) also have antimicrobial activities against fungi, bacteria and even viruses (Lopes et al. 2021). This has been verified for many species belonging to different botanical groups (brown, red or green algae). In the brown alga Fucus evanescens, the total lipid fraction and the different GL subclasses show antimicrobial activity measurable by the diameter of the growth inhibition zone. The antimicrobial activity detected is exerted against fungal pathogens, and also against bacterial strains (see Figure 3.53) (Gerasimenko et al. 2012). This type of activity is, however, less effective than nitrofungin, a disinfectant molecule with a broad spectrum of action (see Figure 3.53).

This antimicrobial activity of algal lipids appears to be highly dependent on the site from which the algae are harvested (see Figure 3.54). Indeed, the lipid fraction of fucus collected from Rudnaya Bay proved ineffective against the fungus *Candida albicans* and the other fungal and bacterial pathogens tested in the study by Gerasimenko et al. (2012). This result differs from those reported for lipid fractions from fucus from other sampling sites (see Figure 3.54). As before, algal lipids are still less effective than nitrofungin (see Figure 3.54).

However, Fucale samples in Rudnaya Bay were collected during a winter campaign (November), when water temperatures tend to fluctuate between 0°C and 1°C. Rockweeds collected at the other sites, on the contrary, came from a summer campaign (August). It is therefore likely that a site effect combined with a seasonal effect was at the root of the variations in biological activity observed.

5 See: https://commons.wikimedia.org/w/index.php?curid=1667576.

Algal species	Lipid type	Type of activity	Active dose (IC_{50})	References
Fucus evanescens (brown alga)	MGDG	Antitumor (melanoma)	100 µM	Lopes et al. (2021)
Sargassum horneri (brown alga)	DGDG SQDG	Antitumoral (colon carcinoma)	100 µM	Lopes et al. (2021)
Sargassum horneri (brown alga)	PL	Antitumoral (leukemia)	Inhibition > 70% at 40 µg/mL	Lopes et al. (2021)
Solieria chordalis (red alga)	MGDG	Antitumoral (Bronchopulmonary carcinoma)	23.5 µg/mL	Kendel et al. (2015)
Gracilaria corticata (red alga)	SQDG	Antitumoral (cervix carcinoma)	106.9 µg/mL	Lopes et al. (2021)
Porphyra crispata (red alga)	SQDG	Antitumoral (liver cancer)	126 µg/mL	Lopes et al. (2021)
Ulva armoricana (green alga)	DGDG	Antitumoral (bronchopulmonary carcinoma)	24 µg/mL	Kendel et al. (2015)
Ulva fasciata (green alga)	Sulfolipids	Antitumoral (lung carcinoma)	0.40–0.67 µg/mL	El-Baz et al. (2013)

Table 3.15. *Some examples of the in vitro antitumor activity of lipids from brown (Phaeophyceae), red (Rhodophyceae) or green (Chlorophyceae) algae*

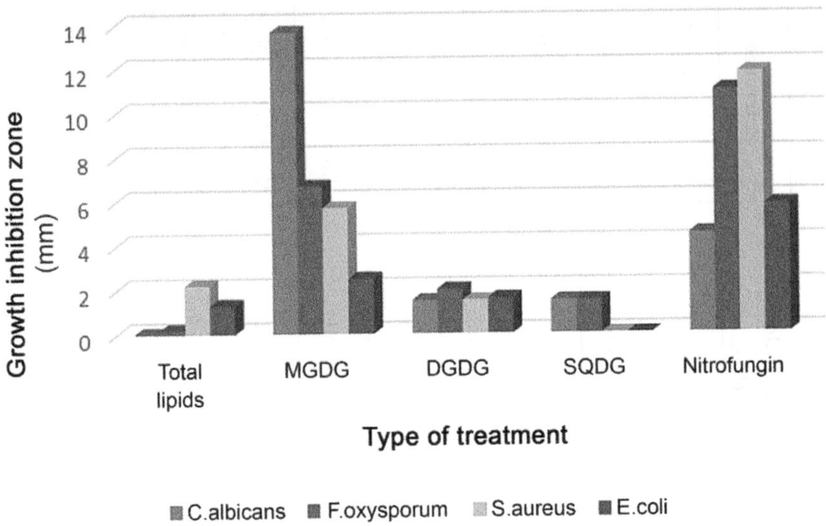

Figure 3.53. *Effect of different lipid classes (total lipids, glycolipids) and nitrofungin on growth inhibition of fungal and bacterial pathogens (according to Gerasimenko et al. (2012))*

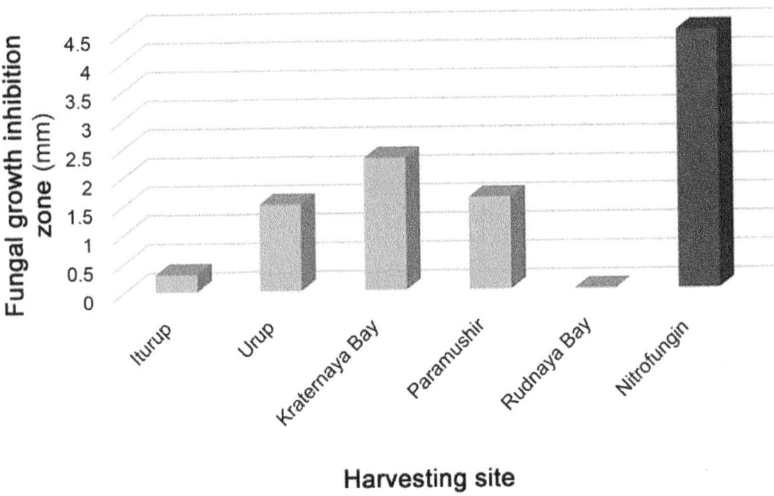

Figure 3.54. *Influence of the sampling site for* Fucus evanescens *on the antifungal activity of the total lipid fraction with regard to* Candida albicans *(according to Gerasimenko et al. (2012))*

In such a context, mastery of the resource – in other words, of the raw material – appears to be an inescapable imperative if the lipid fraction of *F. evanescens* is to be used as a source of antimicrobial molecules (see section 4.2).

When it comes to using GLs, and not the total lipid fraction, the picture is somewhat different. Indeed, the antifungal activity of the various GL subclasses (MGDG, DGDG, SQDG) is greater for fractions obtained from fucus harvested from Rudnaya Bay (see Figure 3.55).

The GL fraction MGDG obtained from samples taken in Rudnaya Bay shows a growth inhibition zone of *C. albicans* of 13.7 mm compared to 4.5 mm with nitrofungin. This study highlights the role played by GLs in antimicrobial activity and more particularly antifungal activity. If the activity of the total lipid fraction is moderate, it is clearly more important with GLs of the MGDG type, for all microorganisms tested (*C. albicans, Fusarium oxysporum, Staphylococcus aureus, Escherichia coli*). This antimicrobial activity seems, as in the previous case, to depend on the geographical habitat, and probably also on the season.

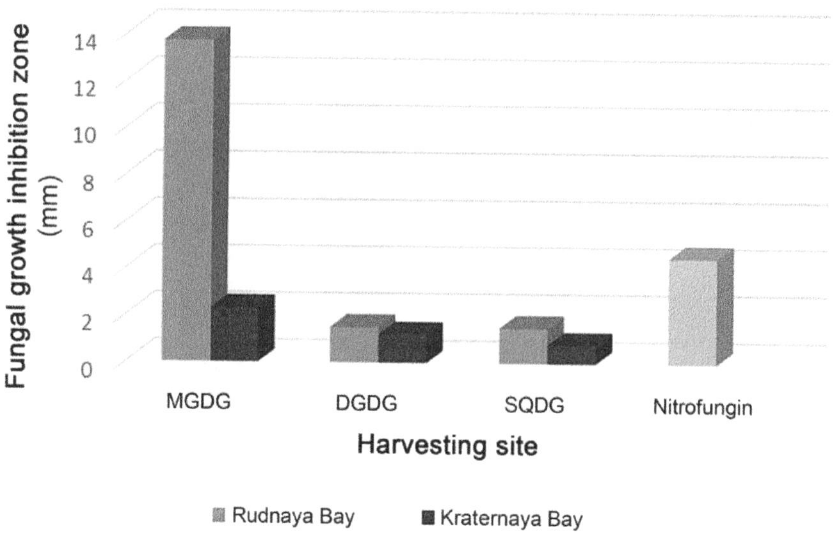

Figure 3.55. *Influence of the sampling site for* Fucus evanescens *on the antifungal activity of different glycolipid fractions (MGDG, DGDG, SQDG) with regard to* Candida albicans *(according to Gerasimenko et al. (2012))*

In addition to antimicrobial activity, lipids from the brown algae *F. evanescens* are also described for their hemolytic activity. This type of activity also appears to be subject to a geographic variation factor (see Figure 3.56). As with antimicrobial activities, GLs are also involved in this activity (see Figure 3.57). Among GLs, the highest hemolytic activity is carried by the DGDG subclass. Indeed, the dose required to obtain a 50% hemolysis rate is 1.55 µg/mL compared to 36.25 µg/mL with SQDG-type GLs (see Figure 3.57).

As with lipids, the fatty acid content of algae is low (1 to 2.8% of dry matter) (Dawczynsk et al. 2007). The content of unsaturated fatty acids of therapeutic interest, such as eicosapentaenoic acid (EPA), varies greatly depending on the algal species (see Figure 3.58). This fatty acid, of the omega-3 type, is one of the protective molecules of the cardiovascular system. In the brown algae *Hizikia fusiformis*, frequently consumed in Japan under the name "Hijiki", EPA represents nearly 42% of the total fatty acid fraction. The consumption of this alga is therefore a complementary source to oily fish in terms of EPA intake. Apart from this type of application, which concerns the preventive rather than the curative field, fatty acids are above all constituents of lipids. They can therefore be associated with the biological activities of the latter, as seen previously.

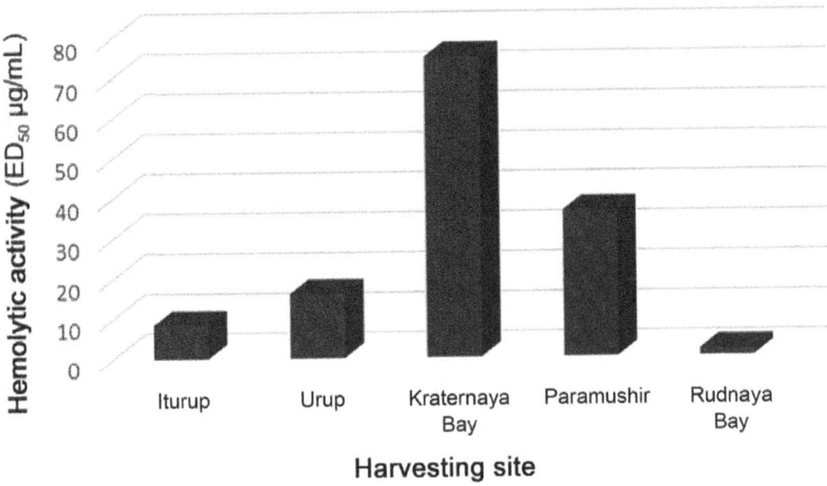

Figure 3.56. *Influence of the sampling site of* Fucus evanescens *on the hemolytic activity of the total lipid fraction (according to Gerasimenko et al. (2012))*

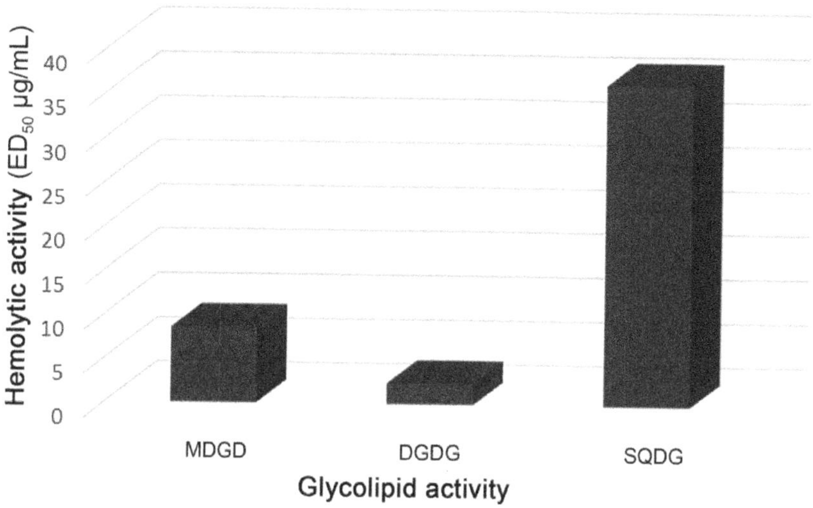

Figure 3.57. *Hemolytic activity of different glycolipid subclasses (MGDG, DGDG, SQDG) derived from* Fucus evanescens *harvested from Rudnaya Bay (according to Gerasimenko et al. (2012))*

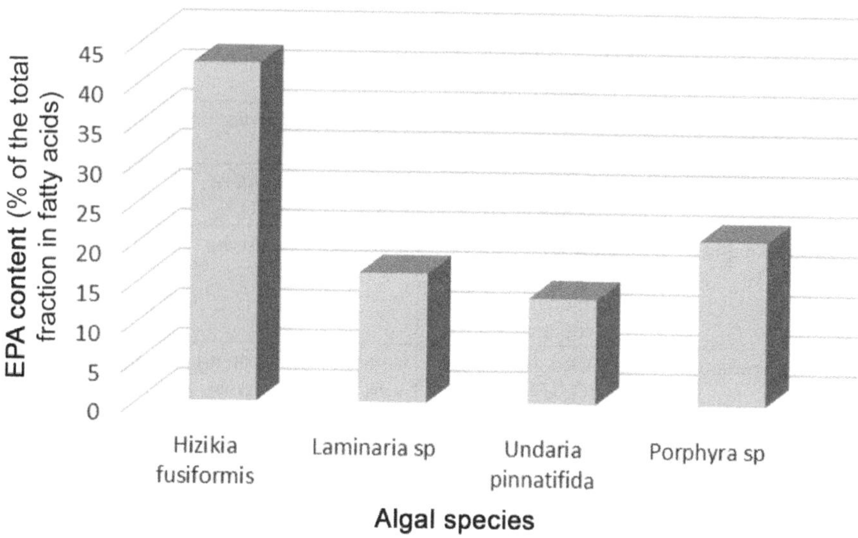

Figure 3.58. *Eicosapentaenoic acid (EPA) content by algal species (according to Dawczynsk et al. (2007))*

Marine algae also contain sterols, for which many biological activities of therapeutic interest have been reported (see Table 3.16). Among these phytosterols, fucosterol is found, which represents 4 to 95% of the sterols present in algae (see Figure 3.49) (Meinita et al. 2021). This compound is mainly present in brown algae (*Fucus* sp., *Eisenia bicyclis, Pelvetia caniculata, Undaria pinnatifida, H. fusiformis, Ecklonia radiata*) and little in green or red algae (Abdul et al. 2016).

Biological activity	Action mechanism	Algal species	References
Antidiabetic	Inhibition of aldose reductase Reduction of insulin resistance	*Eisenia bicyclis* *Ecklonia stonolifera* *Sargassum fusiforme*	Jung et al. (2013a) Abdul et al. (2016) Meinita et al. (2021)
Anticancer	Growth inhibition and induction of apoptosis of cells of the HL 60 line (leukemia) Growth inhibition and induction of apoptosis of cells of the A549, SK-LU-1 lines (lung cancer) Growth inhibition of the P388 line (murine leukemia)	Commercial fucosterol (provided by Sigma-Aldrich) *Sargassum carpophyllum*	Ji et al. (2014) Abdul et al. (2016) Mao et al. (2019) Tang et al. (2002)
Anti-inflammatory	Reduction of free radical production (ROS) and oxidative stress	*Eisenia bicyclis* *Undaria pinnatifida*	Jung et al. (2013b) Meinita et al. (2021)
Antiosteoporosis	Activation of osteoblast proliferation	*Hizikia fusiformis*	Hu et al. (2012)
Antiobesity	Inhibition of adipogenesis	*Ecklonia stonolifera*	Meinita et al. (2021)
Antiphotoaging	Reduction of UVB-induced production of matrix metalloproteinase (collagenase)	*Sargassum fusiforme*	Meinita et al. (2021)
Hepatic protection	Increased glutathione (GSH) level Regulation of cholesterol homeostasis	Commercial fucosterol (provided by Sigma-Aldrich)	Hoang et al. (2012) Meinita et al. (2021)

Table 3.16. *Examples of biological activities of therapeutic interest associated with fucosterol*

In the species *E. radiata*, the blades contain up to 312 µg/g dry weight of fucosterol. This rate increases to 378.1 µg/g dry weight for stipe (Meinita et al. 2021). Fucosterol is described for many biological activities, including antidiabetic, antioxidant, anti-inflammatory, antiosteoporosis and anticancer properties (see Table 3.16).

However, other compounds, derived from fucosterol, can also have biological activities. This is the case for Saringosterol (see Figure 3.59), extracted from the algae *H. fusiformis* and which induces the proliferation of osteoblasts, thus being of interest in the treatment of osteoporosis and periodontitis (Hu et al. 2012).

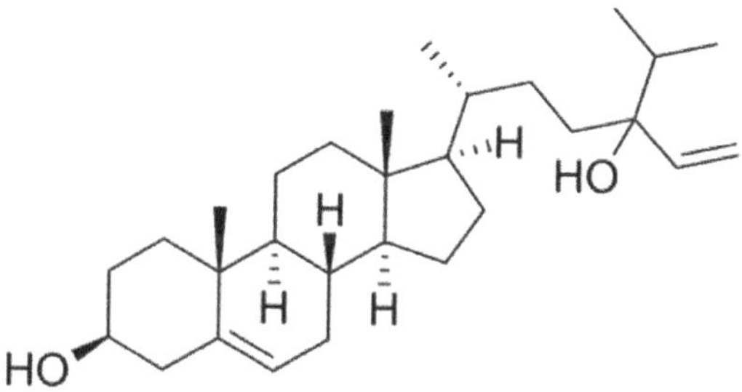

Figure 3.59. *Structure of Saringosterol*[6]

Many studies, however, focus on the antitumor activities of fucosterol. This is mainly due to the reported role of plant sterols as potential agents in the prevention or treatment of cancer. A search using the keywords "phytosterols-cancers" on the Google Scholar search engine reveals 35,000 international publications on the subject.

The study carried out by Mao et al. in 2019 on different lung tumor cell lines shows the efficacy of fucosterol in inhibiting cell growth (see Figure 3.60). Fucosterol is particularly active at 15 µM, inhibiting 50% of tumor proliferation in the A549 and SK-Lu-1 lines.

Algae sterols, and more specifically fucosterol, appear to be interesting molecules for medical research and human health. Furthermore, they are simple molecules, making them easy to synthesize organically. In view of this last point, extracting them from algal resources is not an obvious way of

6 See: https://commons.wikimedia.org/w/index.php?curid=72490778.

adding value. On the contrary, their contribution via the consumption of algae, and Fucales in particular, can be envisaged as a preventive approach to human health.

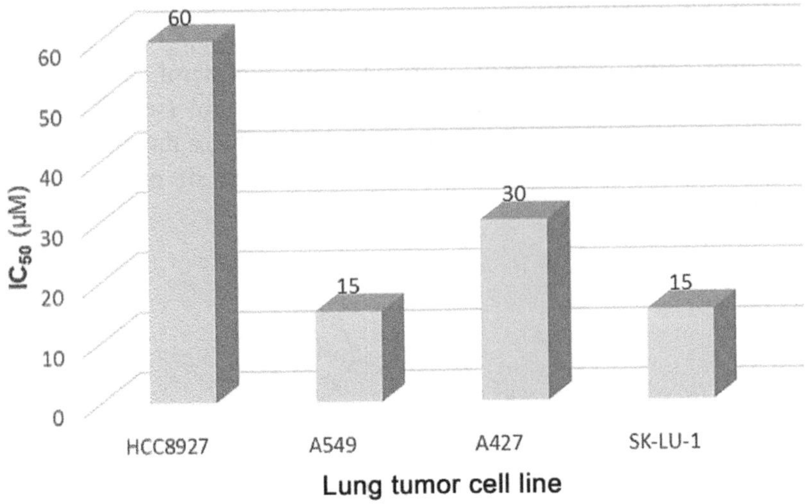

Figure 3.60. *Efficacy of fucosterol on growth inhibition (IC50) of different lung cancer cell lines (according to Mao et al. (2019))*

3.4. Phenolic derivatives

Phenolic derivatives are molecules with an aromatic ring containing one or more hydroxyl groups. They may have a simple structure, such as phloroglucinol, or more complex, such as tetrafucol or phlorotannin (see Figure 3.61). The term "phlorotannins" covers a large number of molecules (dieckol, eckol), whose classification depends on the type of element linked to phloroglucinol. Their content in brown algae is generally between 13 and 15% of dry matter.

Alongside phloroglucinol and its derivatives, algae can also contain brominated phenol derivatives. These are phenol molecules with one or more bromine groups (see Figure 3.62).

Initially isolated from the red alga *Neorhodomela larix*, bromophenols have now been characterized in all taxonomic groups (red, green and brown algae) (Cotas et al. 2020).

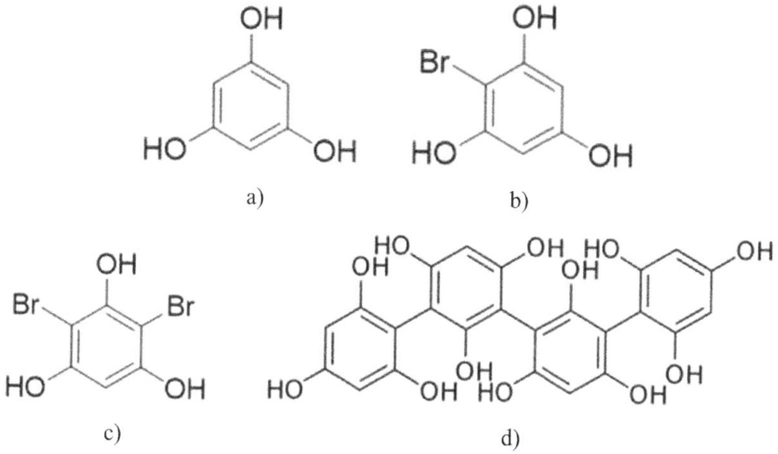

Figure 3.61. *Structure of some phenolic compounds. a) Phloroglucinol; b) bromophloroglucinol; c) dibromophloroglucinol; d) tetrafucol, or pholorotannin, present in the brown alga* Ascophyllum nodosum[7]

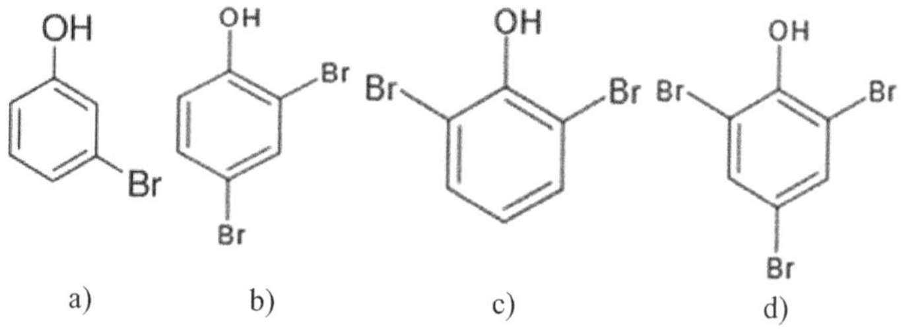

Figure 3.62. *Structure of various bromophenols present in red algae (according to Cotas et al. (2020)). a) 3-bromophenol; b) 2,4-bromophenol; c) 2,6-bromophenol; d) 2,4,6-bromophenol*

Phloroglucinol and its polyphenolic derivatives, namely phlorotannins, are mainly produced by brown algae. They are often associated with antiviral, antitumor, antidiabetic, neuroprotective and anti-inflammatory activities (Cotas et al. 2021).

These molecules are also described for their antioxidant activities, and thus their potentially protective role vis-à-vis carcinogenesis. Nevertheless, in the case of lung cancer lines, the main effects observed are cytotoxic,

7 See: https://commons.wikimedia.org/w/index.php?curid=117659722.

indicating a direct antitumoral action of these compounds (Padua et al. 2015).

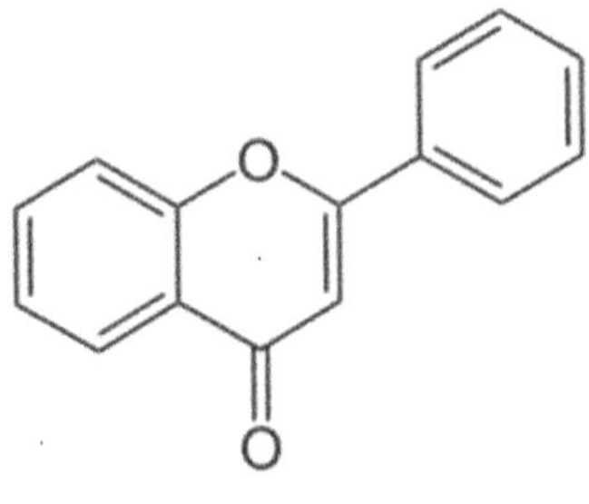

Figure 3.63. *Structure of flavone, one of the main classes of flavonoids present in green algae*[8]

More generally, phenolic derivatives are involved, like other algal compounds, in triggering antitumor mechanisms. The mechanisms involved are cytotoxicity, apoptosis induction, antiproliferative activity and reduction of the risk of carcinogenesis (Cotas et al. 2020).

However, phenolic derivatives, derived from algae such as flavonoids, are often described for their anti-inflammatory and antiallergic activities (see Figure 3.63). A search on the Google Scholar search engine in 2024 revealed 22,800 references published when the keywords phenol-seaweed-anti-inflammatory were entered, compared with 19,300 when the word "anticancer" replaced anti-inflammatory. Adding the references published in the antiallergenic field to those in the anti-inflammatory field led to a total of 31,850 references after cumulating the two fields (see Figure 3.64).

Chronic inflammation is a pre-pathological state with varied but not insignificant consequences for human health (chronic pathologies, allergies). The production of reactive oxygen species (ROS), such as the superoxide ion (O_2^-), plays an important role in triggering and propagating the inflammatory response. The ROS produced behave as chemical signals stimulating inflammatory pathways through the activation of protein kinases, increased genomic expression of pro-inflammatory factors and cytokinin production. Algal polyphenols, such as phlorotannins, which develop strong antioxidant activity, are able to neutralize ROS, thus reducing oxidative stress. The

8 See: https://commons.wikimedia.org/w/index.php?curid=3850694.

presence of aromatic nuclei in the phlorotannin molecule, acting as electron traps, is at the origin of this ROS neutralization mechanism. In particular, these nuclei absorb superoxide and peroxide anions, and hydroxyl radicals from inflammatory oxidation (Besednova et al. 2022). Among the pro-inflammatory compounds of a radical nature is nitric oxide ($NO^{\bullet}$).

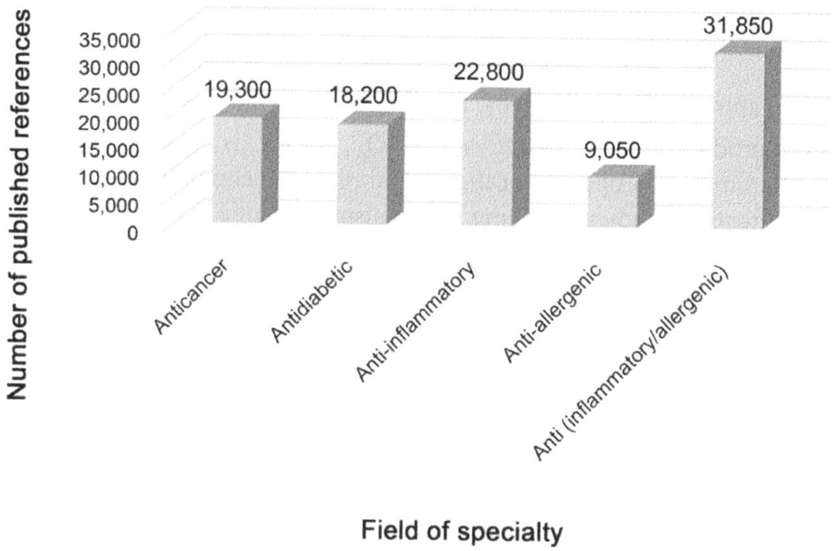

Figure 3.64. *Summary of publications on the various biological activities of phenolic derivatives (according to Google Scholar 2024))*

Added to this are derived molecules such as nitrites, nitrates and nitrosamines. These compounds, when present at excessively high levels, cause severe pathological disorders, such as hypertension, asthma, cardiac dysfunction, septic shock or the triggering of autoimmune diseases.

Phlorotannins from algae of the genus *Fucus* show antioxidant activity towards nitric oxide. However, this activity seems to differ according to the Fucus species considered (see Figure 3.65). Moreover, for a given species, as in the case of *F. vesiculosus*, major differences are observed between harvested and cultivated algae, as well as between algae harvested in different locations (harvested *F. vesiculosus* (1), harvested *F. vesiculosus* (2)) (see Figure 3.65). The highest antioxidant and $NO^{\bullet}$ radical-neutralizing activity was reported for phlorotannins from harvested *F. vesiculosus* (1)

(IC_{50} of 235.9 µg/mL). It was eight times higher than that reported for cultivated *F. vesiculosus* and almost two times higher than that recorded for *Fucus guiryi* (see Figure 3.65).

These results highlight the antioxidant activity of phlorotannins in wrack, and also the variability of the latter, most likely due to the variation in the content of these compounds in the algae depending on the season, site and production method.

This antioxidant activity of phlorotannins, which is exerted on the species produced by oxidative stress (O_2^{-}, $NO^{\bullet}$), is at the root of the anti-inflammatory properties of these algal phenolic compounds. In addition to this direct free-radical trapping mechanism, phenolic derivatives can also modulate the inflammatory response via other mechanisms. In particular, they can prevent activation of the NF-κB protein transcription factor, which plays a key role in the induction of the inflammatory response (Catarino et al. 2020). Through their action on the metabolic signaling pathway of the inflammatory response, phlorotannins can also inhibit the enzymes responsible for NO. production (Cotas et al. 2020).

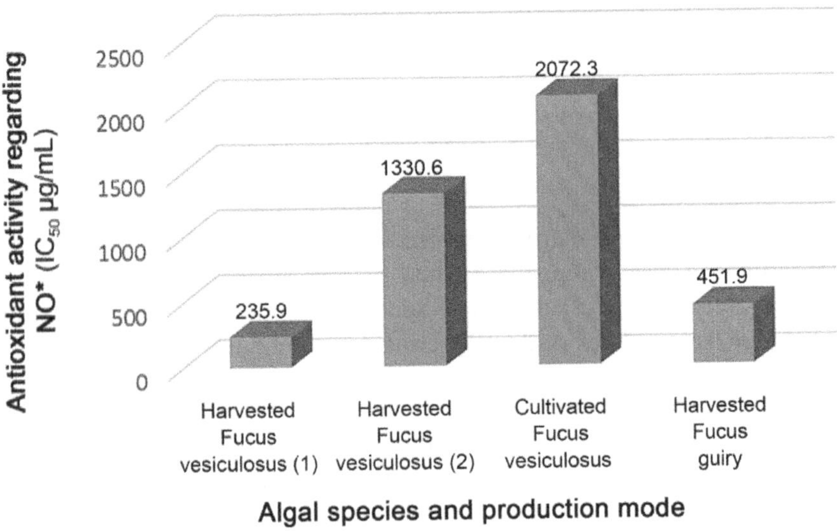

Figure 3.65. *Antioxidant activity of phlorotannins from* Fucus vesiculosus *and* Fucus guiryi *expressed as an inhibition concentration of 50% (IC_{50}) of the oxidizing power of NO species (according to Catarino et al. (2020))*

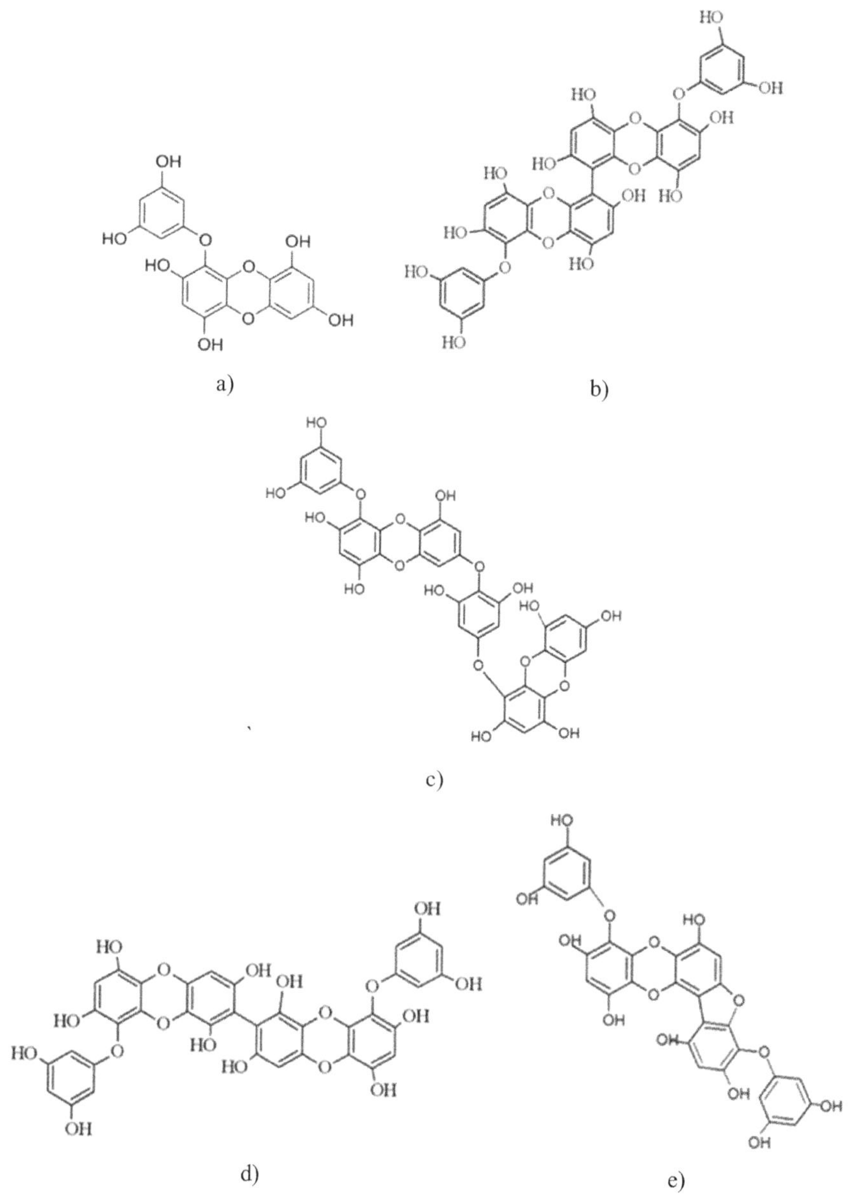

Figure 3.66. *Structure of various molecules belonging to the eckol family. a) Eckol, b) 6,6'-bieckol, c) dieckol, d) 8,8'-bieckol, e) phlorofucofuroeckol (according to Fleurence and Ar Gall (2016[9]))*

9 See: https://commons.wikimedia.org/w/index.php?curid=74546773.

The bromophenols and flavonoids present in green algae are also known for their antioxidant activity. Like phlorotannins, they display free radical trapping activity. They are mainly found in *Ulva clathrata*, *Ulva compressa* (formerly *Enteromorpha compressa*), *U. flexuosa*, *U. intestinalis* (formerly *Enteromorpha intestinalis*) and *Ulva pertusa*.

In addition to antioxidant activity, the phenolic fraction of green algae is also associated with the presence of antitumor activity (Cotas et al. 2020).

Related to anti-inflammatory activity, phenolic compounds, such as phlorotannins, are also described for their antiallergenic activity (Fleurence and Ar Gall 2016). These molecules belong to the eckol family and are mainly present in brown algae (see Figure 3.66). More specifically, these are the species *Ecklonia kurome*, *Ecklonia stolonifera*, *E. bicyclis*, *Eisenia hyperborea* and *Eisenia nipponica*, which are eaten as sea vegetables in Korea and Japan. The antiallergenic mechanisms of eckols have been partially elucidated in both in vitro (RBL-2H3 cells) and in vivo (murine) models (Sugiura et al. 2021).

In particular, they inhibit basophil degranulation (RBL-2H3), thus preventing the release of histamine from cytoplasmic granules. Inhibiting the release of this chemical mediator blocks the allergic reaction and avoids the associated symptoms (urticaria, gastric bloating, respiratory discomfort).

However, different eckol molecules have different efficacy in inhibiting basophil degranulation. The highest histamine release inhibition activity is found with phlorofucofuroeckol (IC_{50} 4.3 μM), which remains a structurally complex eckol (see Figures 3.66 and 3.67).

The antidegranulation activity of eckols is not limited to the suppression mechanism of histamine release by basophils. In fact, 6,6'bieckol, dieckol and phlorofucofuroeckol are also involved in inhibiting the binding of IgE and FcεR1 antibodies to basophils and mast cells. This adhesion is at the origin of the cell activation mechanism, which is an important step in triggering one of the many pathways of the inflammatory and allergic process.

In addition to mechanisms linked to the production of free histamine, the eckol family also appears to be involved in the attenuation of inflammatory and allergenic metabolic pathways, such as the MAPK (mitogen-activated protein kinases) pathway and the inhibition of phosphorylation of transcription factors playing a signaling role in this pathway (NF-κB).

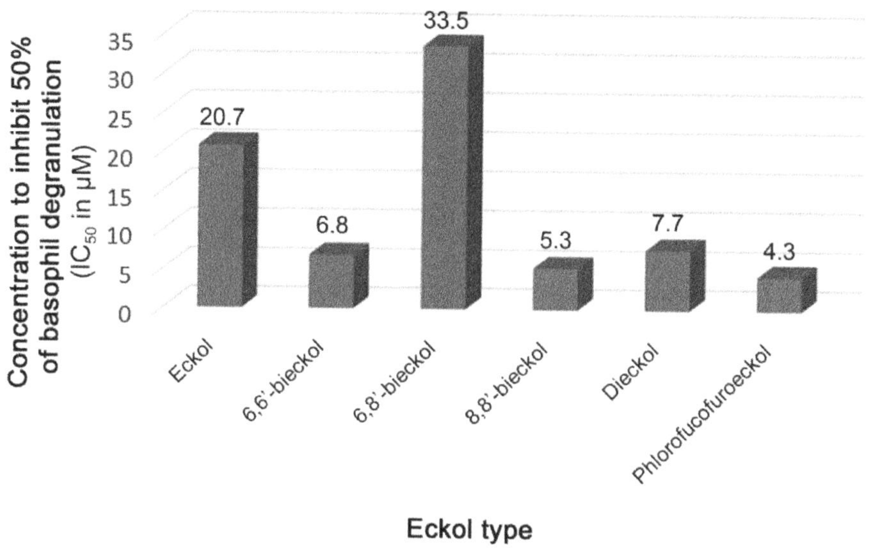

Figure 3.67. *Comparative inhibition (IC_{50}) of different eckol molecules regarding degranulation of RBL-2H3 basophils (according to Sugiura et al. (2021))*

Eckols and their related derivatives also play a complex immunomodulatory role, notably by suppressing the production of IgE antibodies required for the cellular activation of histamine-producing basophils.

This set of results explains the limited number of reported cases of food allergies linked to algae consumption (James et al. 2023; Fleurence 2024).

Algae contain macromolecules with interesting activities for human health. These include polysaccharides (alginates, laminarin, fucans, carrageenans and ulvans). These constituents are present in large quantities in algae, and the extraction processes applied to them are well mastered. On the other hand, their molecular weight and their capacity to generate viscosity in aqueous media represent a real disadvantage for intravenous administration. Oral administration is possible, and certain sodium alginate-based products are already marketed as antireflux drugs. The same type of alginate is also used in hemostatic dressings, which are applied to the skin.

Apart from polysaccharides, whose mode of administration and toxicity limit their therapeutic application, their oligosaccharide derivatives, with lower molecular masses, are better candidates for drug development.

The case of proteins is more nuanced. In fact, these constituents are present in more moderate quantities than polysaccharides. This general observation does not apply to red algae, whose protein content can reach up to 47% of dry matter in some species. It is further qualified by the presence of phycobiliproteins and, more particularly, R-phycoerythrin, which have biological activities of pharmaceutical interest. These algae-specific oligomeric proteins cannot be produced by synthesis, hemisynthesis or genetic engineering. Their extraction from algal resources therefore remains the only possible method.

Other algal constituents of pharmacological interest (peptides, lipids, sterols, phenols and derivatives) are present in smaller quantities and are sometimes subject to significant seasonal variations. In the latter case, algae do not appear as an interesting resource for industrial-scale extraction of these types of molecules. On the contrary, these molecules, which are easy to synthesize by chemical, enzymatic or biotechnological means, can be considered as potential models for the development of new drug molecules.

4

Therapeutic Development: Strategies and Limits

The use of algae in human health can be conceived in two ways. The first is to incorporate algae into the diet, in order to benefit from the preventive effects of this diet on health. The second is to ensure, on a curative basis, the development of new drugs derived from algae. This dual strategy is nevertheless subject to many limitations.

4.1. Epidemiological studies

The beneficial effects of algae consumption on human health have been the subject of a number of epidemiological studies (Brown et al. 2014).

The Japanese island of Okinawa is world-famous for its high number of centenarians. The diet of this population consists mainly of fish, soy and algae.

For the Japanese archipelago as a whole, the daily algae consumption is estimated at 5.3 g.

A study carried out on a cohort of 40,547 men and women aged between 40 and 79 showed that the traditional Japanese diet induced a lower cardiovascular risk, compared to a Western diet rich in animal proteins (Shimazu et al. 2007). The Asian diet includes seaweed, a sea vegetable traditionally consumed in Japan and Korea.

A more limited study, carried out on some 20 people of Japanese origin, whose ancestors had migrated to Brazil, also showed the beneficial effect of a diet incorporating algae (5 g/day) on reducing blood cholesterol levels, the main cardiovascular risk factor (Yamori et al. 2008).

An epidemiological study carried out in Korea on a cohort of 362 women showed that the risk of breast cancer was inversely proportional to the consumption of the algae *Porphyra* sp. (nori or gim in Korean) in dehydrated form in premenopausal individuals (Brown et al. 2014). On the other hand, no correlation was observed in the case of dietary intake of *Undaria pinnatifida* (wakame or miyeok in Korean) (Yang et al. 2010).

In contrast, the same type of study carried out on a population of 34,759 Japanese women showed no such correlation for nori. Nonetheless, the latter study shows some bias in the way it was carried out. It did not specify to consumers the search for a correlation between the consumption of sea vegetables and breast cancer. Information was collected not by interview, but by e-mail response to questionnaires, thus limiting the verification of responses.

A collaborative Japanese study also found that the dietary intake of algae was associated with lower lung cancer mortality in both men and women. Pancreatic cancer mortality also appeared to be lower for men including algae in their diet (Iso and Kubota 2007; Brown et al. 2014). The protective effect of algae is most notable in the case of lung cancer (Iso and Kubota 2007). In contrast, an epidemiological study involving 18,115 Japanese showed no correlation between algae consumption and a protective effect against prostate cancer (Allen et al. 2004). In this study, the consumption of fish more than four times a week significantly increased the risk of prostate cancer. This result contradicts that of the Iso and Kubota study (Iso and Kubota 2007), which shows that fish consumption has a protective effect. This kind of finding shows how difficult it can be to draw conclusions from certain studies.

Many epidemiological studies based on large cohorts rely on questionnaires relating to the dietary habits of interviewed individuals, and on the veracity of the answers given.

On the other hand, other studies, based on smaller numbers, rely on a better mastered scientific methodology. This is the case of the study on the protective effect against breast cancer when consuming the brown alga

Undaria pinnatifida (wakame) (Teas et al. 2013). This work was carried out on 15 postmenopausal patients who had or had not experienced breast cancer. Patients were divided into three groups (no diagnosed breast cancer or control group, breast cancer with estrogen receptor ER+, breast cancer without estrogen receptor ER–). The 15 patients were also defined according to social, economic, demographic and age criteria (see Table 4.1). Patients were placed on a diet in which they successively consumed 10 daily capsules (5 g/d) of placebo 1 for four weeks, *Undaria* for four weeks and placebo 2 for four weeks. Urine analysis of the plasminogen-urokinase activator receptor (uPAR) was used to assess treatment efficacy. This molecule is involved in cell adhesion and the detachment of cancer cells from the extracellular matrix (ECM).

As shown, a very low dietary intake of algae (5 g/day) resulted in a very significant reduction in uPAR in patients who had or had previously had breast cancer, whatever the group (ER+ or ER–) (see Figure 4.1).

This protective effect of the *U. pinnatifida* consumption in postmenopausal women was not confirmed by the study carried out by Yang et al. (2010) on premenopausal Korean women.

The inclusion of algae in the diet of Japanese women could at times explain the lower incidence of breast cancer mortality in this population compared to that recorded for Western women.

	Control group	**ER+ group**	**ER– group**
Age (years)	5 (56–64)	5 (55–66)	5 (50–66)
African-American	2	0	4
European	3	5	1
High school	1	0	1
Junior high	1	2	1
Bachelor's degree or higher	3	3	3
Income			
$20,000–49,000	2	1	1
$50,000–99,000	2	1	3
> $100,000	1	2	0
No information	0	1	1

Table 4.1. *Demographic and social–economic profiles of the sample of patients selected for the study (according to Teas et al. (2013))*

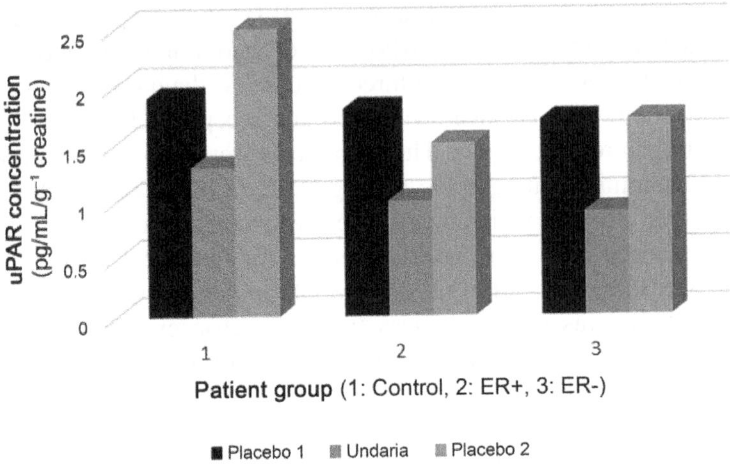

Figure 4.1. *Effect of the oral administration of Undaria pinnatifida on the urinary concentration of the marker and tumor inducer uPAR (according to Teas et al. (2013)). For a color version of this figure, see www.iste.co.uk/fleurence/health.zip*

Algae, particularly brown algae, contain high levels of iodine. Frequently eaten algae such as *Porphyra* (nori), *Undaria* (wakame) or *Laminaria* (kombu), have iodine levels ranging from 80 to 5,000 mg/kg dry matter. Japanese consumers ingest up to 12 mg of iodine per day, while Western consumers consume between 166 µg and 240 µg per day. Iodine intake is sometimes thought to have a protective effect against breast cancer. Nevertheless, few epidemiological studies demonstrate this type of correlation.

On the other hand, iodine is suspected of being a risk factor for the development of thyroid cancer. A study carried out on a cohort of 52,679 Japanese women showed a strong correlation between algae consumption and the increased risk of thyroid cancer (Michikawa et al. 2012). The study was carried out over nearly 15 years on women aged between 40 and 69. Their algae consumption habits were divided into three groups:

– group 1, consumption two days a week or less;

– group 2, consumption three to four days a week;

– group 3, daily consumption.

Daily consumption of algae, and thus excessive iodine intake, significantly increased the risk of thyroid cancer (papillary carcinoma). This risk was particularly marked in postmenopausal women, and much less so in premenopausal women.

This kind of study, like those on breast cancer, shows that the protective effect or otherwise of an algae-based diet is also associated with the hormonal status of patients.

Epidemiological studies on the impact of algae consumption in the prevention of chronic pathologies can sometimes generate contradictory results. This is often due to methodological bias (questionnaires validated in an insufficiently controlled manner, poorly formulated questions, cohort sizes too small). In the case of certain cancers (breast cancer, thyroid cancer), the reduced production of estrogenic hormones appears to have a direct influence on the protective effect of algae consumption.

Algae contain numerous anti-cancer molecules (polysaccharides, phycobiliproteins, glycolipids) whose in vitro activity on cell lines and in vivo activity on murine models are well known.

Against this backdrop, it is reasonable to assume that the consumption of algae could have a positive impact on human health in terms of prevention. Several epidemiological studies also point in this direction.

Epidemiological studies, their methodology and the complex nature of the factors taken into account (dietary diversity, social and economic criteria) are interesting tools, but in some ways limit the promotion of algae consumption in terms of human health prevention. To this point, we must add the historical and cultural dimension of the limited acceptability of algae as a traditional food by the Western consumer.

4.2. Clinical studies

Among the clinical studies carried out on the administration of algae or their molecules as part of the treatment of a pathology, a distinction must be made between public and private approaches.

In the first case, results are published in the scientific community and are therefore easily accessible. In the second case, publication of results is more confidential and limited to a panel of experts, industrial decision-makers and health agencies.

Polysaccharides	Mechanisms of action	Consequences for MetS
Alginates	Inhibition of digestive enzymes (pancreatic lipase, pepsin) Faster filling of the gastric bolus	Reduced intestinal absorption of triglycerides, cholesterol and glucose Reduces food intake by reaching satiety more quickly
Fucoidans	Angiotensin-converting enzyme inhibition Inhibition of α-amylase and α-glucosidase	Lower blood pressure Decreased intestinal glucose absorption
Laminarin	Reduced intestinal absorption of cholesterol Lower systolic pressure	Reduced levels of lipids and cholesterol levels Lower blood pressure

Table 4.2. *Examples of the mechanisms of action of the main polysaccharides of brown algae on metabolic syndrome (MetS) (according to Keleszade et al. (2021))*

Among the published and easily accessible clinical studies, there is a work, published in 2021, on the clinical efficacy of the administration of brown algae extracts in the prevention of metabolic syndrome (MetS). This syndrome is not a disease in itself, but it is a precursor of more serious pathologies, such as obesity, type II diabetes and cardiovascular accidents. Almost 26% of the world's population seems to be affected by this syndrome, which has become a global public health issue (Keleszade et al. 2021). The brown algae *Ascophyllum nodosum* and *Fucus vesiculosus* contain various polysaccharides (alginates, fucoidans and laminarin) known for their activity on lipid metabolism, carbohydrate metabolism and hypotensive action (see Table 4.2).

In most clinical studies, the dietary intake of algae is via capsules or food containing an algal ingredient.

A Canadian study carried out on 33 patients (11 men, 12 women) aged between 19 and 59 shows the effect of assimilating *Ascophyllum nodosum* or *Fucus vesiculosus* on plasma glucose and insulin concentrations (Paradis et al. 2011). Consumption of 500 mg of algae (two 250 mg capsules) 30 minutes before eating 50 g of white bread carbohydrates had an impact on blood insulin concentration. The latter was reduced by 12.1% compared with a placebo.

Another study, carried out on a sample group of 43 people (19 men, 24 women) aged between 21 and 63, in good health or at cardiovascular risk, also shows the impact of algae consumption on certain aspects of the metabolic syndrome (Iacoviello et al. 2013). The algae were supplied in the form of commercial capsules containing 900 mg of *A. nodosum* powder and 175 µg of iodine. After six weeks, the algae supplementation group saw a significant decrease in the weight of study participants. This weight reduction was also observed in the placebo control group. On the other hand, a significant decrease (–5%) in glyceride levels was observed in members of the algae diet group. This reduction was likely to reduce metabolic syndrome, and hence cardiovascular risk.

A study carried out on a cohort of 175 patients with type II diabetes (85 men, 90 women) and overweight (BMI) shows a positive effect of algae supplementation on the reduction of abdominal circumference (Derosa et al. 2019). The supplement consisted of three daily capsules, each containing 237.5 mg of *Ascophyllum nodosum*, 12.5 mg of *Fucus vesiculosus* and

7.5 μg of chromium picolinate. This intake resulted in a reduction in carbohydrate metabolism parameters such as glycemic hemoglobin levels and various plasma glucose forms (FPG, PPG). On the other hand, the diet had no effect on lipid metabolism parameters (HDL or LDL cholesterol, triglycerides).

Another study, published in 2018 and carried out on patients suffering from symptoms of MetS, describes comparable effects (De Martin et al. 2018). This work was based on a cohort of 50 volunteers (18 men, 32 women). Eighteen patients suffered from hypertension, eight from thyroid dysfunction and six from obesity. The supplementation with algae (*Ascophyllum nodosum, Fucus vesiculosus*) significantly reduced abdominal circumference after six months of treatment. Abdominal circumference fell from 112 cm+/-17 to 105 cm +/-13. Glucose and insulin levels were also reduced after six months of treatment (see Figure 4.2). This is particularly true when calculating the HOMA index, which is the fasting glucose/insulin ratio. When this ratio is low, it indicates insulin sensitivity, and therefore a non-diabetic state. In the case of algae supplementation, a decrease in this ratio was noted from the third month onwards, continuing until the sixth month (see Figure 4.2).

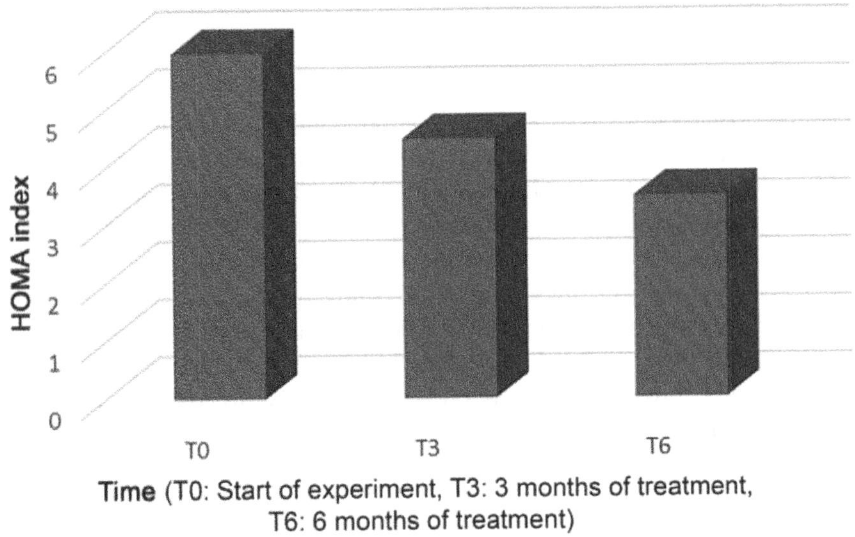

Figure 4.2. *Effect of algae supplementation* (Ascophyllum nodosum, Fucus vesiculosus) *on the HOMA index of volunteers (according to De Martin et al. (2018))*

A dozen clinical studies on the impact of algae supplementation on the main factors of metabolic syndrome have been published since 2011 (Keleszade et al. 2021). As seen above, they concern the inclusion in the diet of two species of brown algae, *Ascophyllum nodosum* and *Fucus vesiculosus*. The effects obtained on MetS parameters differ according to the medical profile of the volunteers included in the initial cohorts (e.g. diabetic or not). In some cases, an impact on patients' lipid profile was noted, while in others it was not. The same applies to the carbohydrate profile, which may or may not be affected by algae supplementation.

Generally speaking, the main effects described in these studies concern a reduction in abdominal circumference and better glycemic control. In some, but fewer cases, an impact on lipid levels was also observed.

Clinical studies on the positive effects of algae on human health mainly concern nutrition and disorders resulting from dietary imbalance (type II diabetes, hyperlipidemia, obesity). These studies are more preventive than curative.

In the curative field, in situ studies on the effect of algal molecules on pathologies are of course carried out on animal species, including mammals. However, these studies are rarely made public, and the data are not easily accessible. The same applies, of course, when the clinical studies concern the human model.

4.3. Patents

The biological activities of algal polysaccharides (alginates, carrageenans, fucans, porphyrans, ulvans) are the subject of a large number of patents worldwide (see Figure 4.3). The claims developed concern the main pharmacological properties mentioned above, as well as those linked to the development of new biomaterials for reconstructive surgery. For alginates and carrageenans, the number of patents issued exceeds 100,000, according to the Google Patents database. For these two polysaccharides, the patents issued cover a very wide range of applications related to their biological activities, as well as to their physicochemical properties, which could be used in the formulation of new drugs.

Fucans, polysaccharides derived from brown algae, have been the subject of a more limited number of patents (see Figure 4.3). These are mainly associated with biological properties such as anticoagulant, antithrombotic, immunomodulating or antiviral activity.

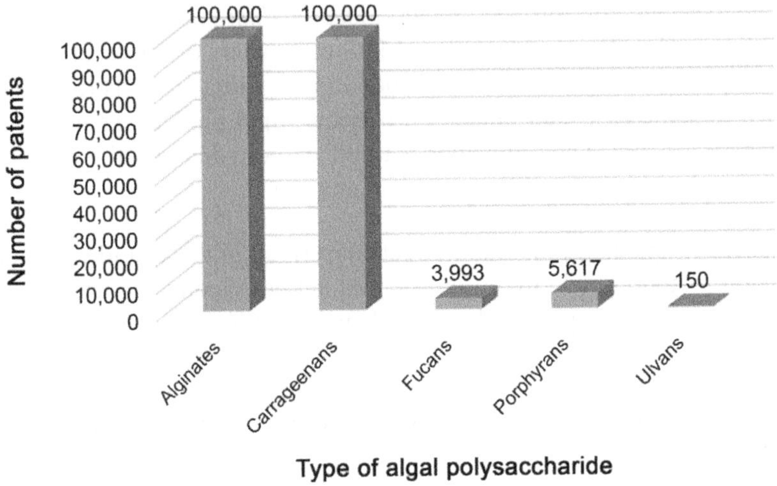

Figure 4.3. *Number of patents concerning the biological activities of algal polysaccharides according to the Google Patents database (March 2024)*

Since the end of the 1980s, several patent applications on the use of fucans as potential anticoagulants have been filed at the French and European level.

The first patent concerns the use of saccharide derivatives obtained by controlled hydrolysis from fucans extracted from brown algae and their use as new anticoagulant, antithrombotic or anticomplementary active agents. The oligofucans obtained are characterized by molecular weights of between 10 and 40 kDa, making them easier to administer than the original fucans. This patent is interesting because it combines two potentially protectable aspects.

The first concerns the development of controlled hydrolysis processes for fucans extracted from algae. The processes involved are many and varied, and include chemical hydrolysis (H_2SO_4 0.5-1N), radiolysis (γ-radiation), sonication or enzymatic hydrolysis (unspecified enzyme). The second aspect concerns the fields of application of these fucan-derived molecules, mainly anticoagulant activities[1].

1 See: https://patentimages.storage.googleapis.com/b6/10/5c/cd64c79ea96134/EP0403377B1.pdf.

In 2001, another patent was issued to complement the previous one. It relates to the development of a low-molecular-weight fucan (10,000 g/mol) potentially active against thrombosis and arterial restenosis[2].

In 2014, a patent on the use of aminofucoidans was filed by a French team. This synthetic molecule, which is involved in activating fibrinolysis of the thrombotic clot, is part of a strategy to develop new antithrombotic drugs[3]. In addition to the activities classically reported for fucans, the use of fucans in tissue regeneration has also been patented. The claim relates to the use of fucans in the activation of dermal fibroblasts, and thus in the reconstruction of connective[4] tissues.

Finally, and more anecdotally, a Chinese patent issued in 2007 claims the use of fucans in the prevention and treatment of Parkinson's disease[5].

The number of patents protecting the use of polysaccharides and other algal molecules in the field of human health is very high. However, there are a very few active ingredients of algal origin in today's medicines, with the exception of the use of alginates in certain pharmaceutical specialties (gastric or hemostatic dressings, biomaterials).

This limitation of the development of algal assets in the field of pharmacopoeia has several origins. These include accessibility, availability and biochemical variability.

4.4. Resource availability, accessibility and biochemical variations

The use of algae and its compounds in human health can be considered from a preventive and curative perspective. From a preventive point of view, this necessarily means incorporating more algae into our diets, especially in Western countries.

---

2 See: https://patentimages.storage.googleapis.com/57/de/38/8f0f39f48b58e1/WO2001015654A2.pdf.

3 See: https://patentimages.storage.googleapis.com/7b/f2/91/8845e5b41ffd33/WO2014147597A1.pdf.

4 See: https://patentimages.storage.googleapis.com/9d/f9/da/c39779f6e47b0d/WO1999032099A2.pdf.

5 See: https://patentimages.storage.googleapis.com/40/71/d8/f43e934d3af199/CN101301310B.pdf.

In the curative field, certain algal species, mainly brown algae, have already been integrated into the therapeutic practice of traditional Asian medicines.

On the other hand, extracting molecules for medicinal purposes from algal biomass poses other problems.

The first concerns the availability of the resource in terms of tonnage and ease of recovery. The second involves controlling the biochemical composition of the resource and eliminating excessive seasonal variations.

The solution to these two problems lies in cultivation or algoculture. To date, there are only a limited number of algae cultivated on a large scale and capable of paving the way for a more or less standardized supply to industry. Nevertheless, for the past two decades, the global production of algae by aquaculture has been steadily increasing (see Figure 4.4).

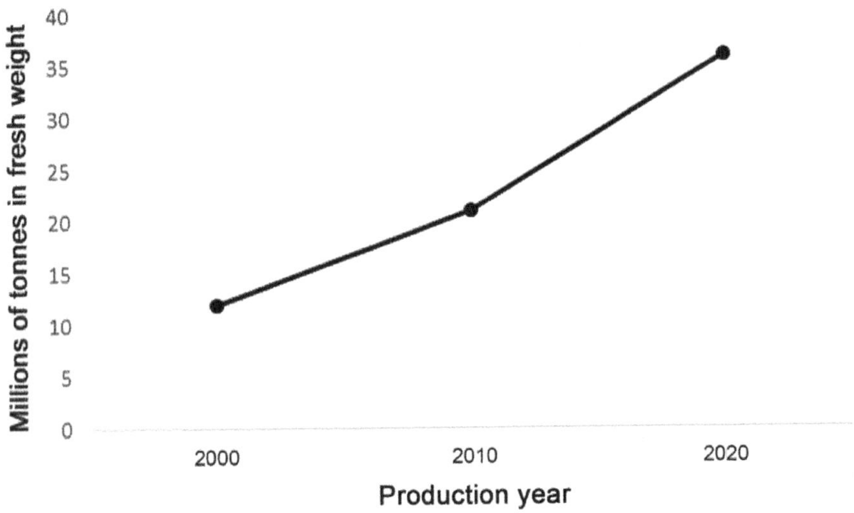

Figure 4.4. *Evolution of world algae production by crop over the last two decades*[6]

6 See: https://www.fao.org/3/cc0461fr/online/sofia/2022/world-fisheries-aquaculture-production.html.

Today, algae farming accounts for 97% of the world's algae production, but is mainly limited to a few species of brown algae (*Laminaria japonica*, recently *Saccharina japonica*, *Undaria pinnatifida*) and red algae (*Porphyra* sp., recently *Pyropia*, *Kappaphycus alvarezii*, *Eucheuma denticulatum*).

Many species, used in traditional medicine or described for the pharmacological activities of their molecules, are not available through large-scale cultivation. Although they can be harvested, this method does not allow us to control the resource in quantitative, qualitative or economic terms.

The world's algae production is mainly located in Asia, with China as the world's leading producer, which has mainly focused on growing *Laminaria japonica* to meet the needs of the alginate and food markets (see Figure 4.5).

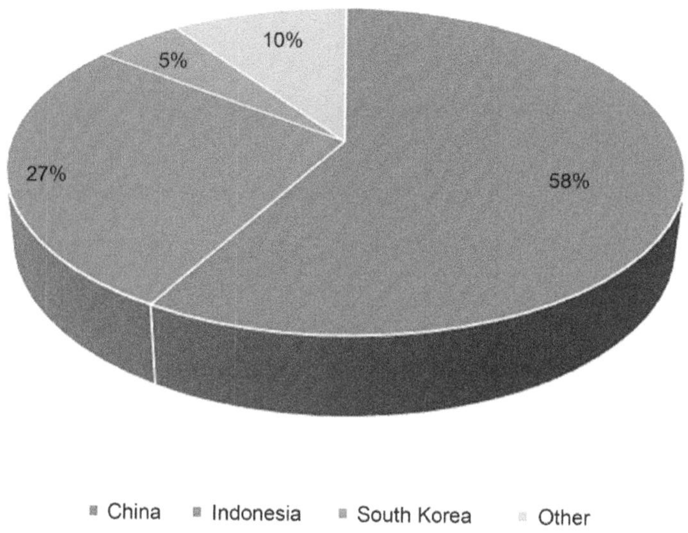

Figure 4.5. *Global algae production by world region[7]. For a color version of this figure, see www.iste.co.uk/fleurence/health.zip*

This state of affairs creates a dependency on the accessibility of the resource in the case of the development of new applications for algae in the field of human health.

7 See: https://www.fao.org/3/cc0461fr/online/sofia/2022/world-fisheries-aquaculture-production.html.

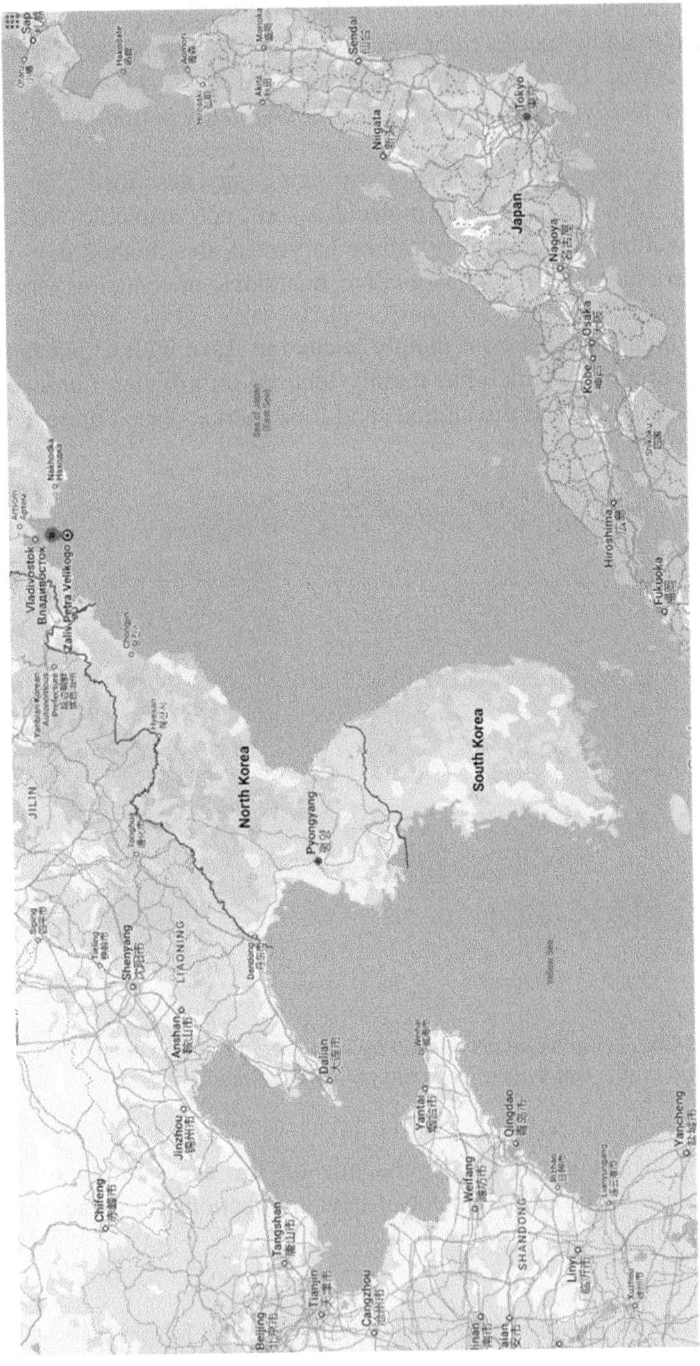

Figure 4.6. *Location of the Peter the Great Gulf in the Sea of Japan (from Google Maps). For a color version of this figure, see www.iste.co.uk/fleurence/health.zip*

Among the molecules most frequently cited in the literature for their therapeutic activities are algal polysaccharides. The latter, also known as "phycocolloids" (alginates, agar, carrageenans, fucans, laminarin), are characteristic of algae. The algal biomass remains the only biological resource capable of supplying these molecules.

Phycocolloid contents vary according to the species, season and even geographical location. This is true for fucans, and more specifically for fucoidans, which have numerous biological activities of pharmaceutical interest (see section 3.1.3).

In fact, in the case of *Saccharina japonica* (*Laminaria japonica*), harvested in Peter the Great Gulf (Sea of Japan), the level of fucoidans varies from 0.98 to 4.19% of dry matter over the course of the year (see Figure 4.6) (Skriptsova 2016).

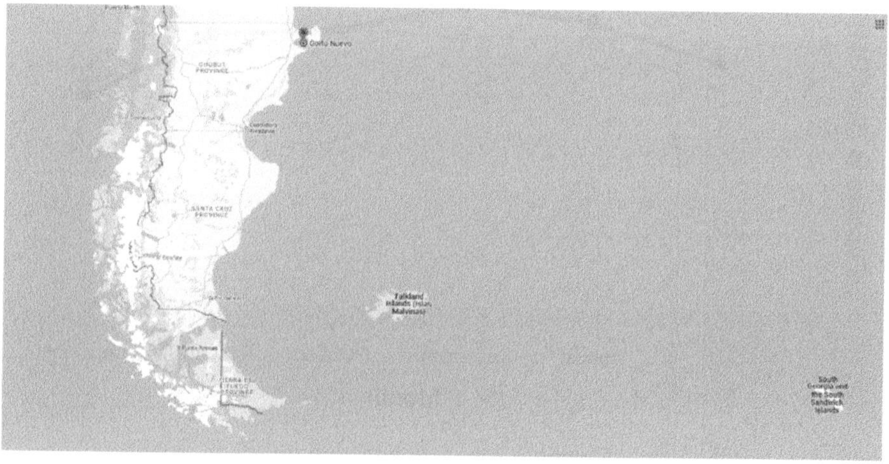

Figure 4.7. *Location of Golfo Nuevo, Patagonia*[8]. *For a color version of this figure, see www.iste.co.uk/fleurence/health.zip*

In another laminariale species, *Undaria pinnatifida*, the fucoidan content is significantly higher than that reported for *Saccharina japonica*. It varies between 14.5 and 19.2% of dry weight (Arijon et al. 2021). Sampling was

8 See: https://www.google.fr/maps/place/Golfe+Nuevo/@-42.7323217,-64.9202631,107118m/data=!3m2!1e3!4b1!4m6!3m5!1s0xbe030084572ddf19:0xfcbbd88ecad08325!8m2!3d-42.7375186!4d-64.4787854!16s%2Fm%2F0420xkf?entry=ttu.

carried out not in the Sea of Japan, but in Argentina's Golfo Nuevo (see Figure 4.7). This site is a protected bay, less open than Peter the Great Gulf, which is exposed to the currentology and temperatures of the Sea of Japan.

In brown algae of the genus *Sargassum,* the impact of season and geographical location on the fucoidan content is well described (Rani et al. 2017). For the species *Sargassum wightii*, a variation in the fucoidan content is observed during the different seasons, as well as according to the harvesting location (see Figure 4.8). The two sites chosen for this study are located in India, in the Gulf of Mannar, part of a wide inlet between India and Sri Lanka (see Figure 4.9). The Hare Island site is characterized by a shallow depth, and the Valinokkam area by a higher bathymetry. This parameter, combined with geographical location, is also a factor in inducing variations.

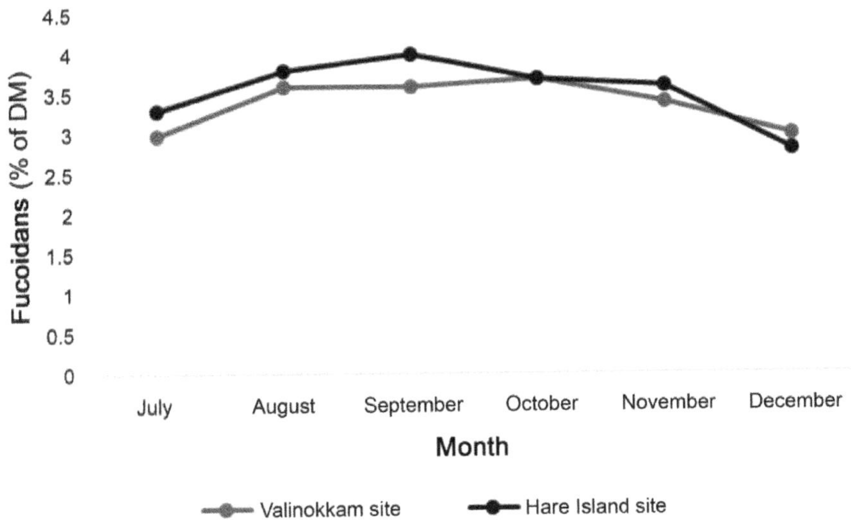

Figure 4.8. *Variation in the fucoidan content in the brown alga* Sargassum wightii *according to season and geographical sampling site (according to Rani et al. (2017)). For a color version of this figure, see www.iste.co.uk/fleurence/health.zip*

In the field of algal macromolecules other than polysaccharides with potential for pharmaceutical development are phycobiliproteins such as R-phycoerythrin (R-PE). The content of this protein pigment also varies according to species, season and sampling site.

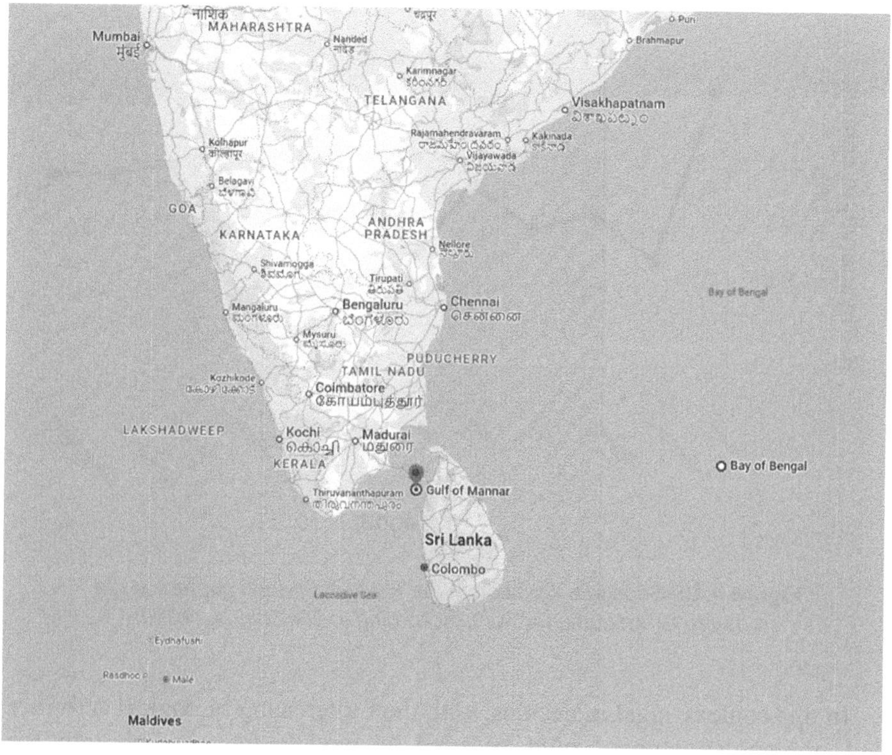

Figure 4.9. *Location of the Gulf of Mannar*[9]*. For a color version of this figure, see www.iste.co.uk/fleurence/health.zip*

The species *Grateloupia turuturu* show a strong seasonal variation in R-PE content. It is highest in January and lowest in August (see Figure 4.10) (Denis et al. 2010).

The use of algae as a source of molecules for therapeutic purposes is therefore limited by all the factors mentioned above. As in the case of many plant sources, resource availability remains the primary limitation. In the case of algae, the main cultivated species, *Saccharina japonica*, is monopolized by the global alginate and agri-food markets, and the food industry. The other cultivated species are used extensively for the production of other phycocolloids (agar, carrageenans) and sea vegetables.

9 See: https://www.google.fr/maps/search/Gulf+of+Mannar+Hare/@9.9100864,78.1365683, 1149245m/data=!3m1!1e3?entry=ttu.

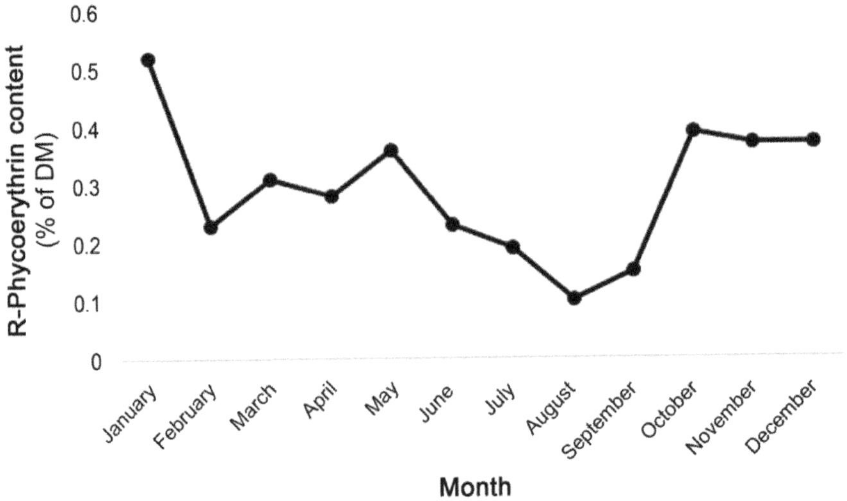

Figure 4.10. *Seasonal variation in the R-phycoerythrin content in the red alga* Grateloupia turuturu *(according to Denis et al. (2010))*

In this context, algal molecules, with their interesting biological activities and original structures, could be used as models for research into the development of new therapeutic agents. Their production could be based on organic synthesis or hemisynthesis for the simplest molecules, and genetic engineering for the most complex.

4.5. Perception and ignorance of the resource

Algae are relatively unknown in the Western world. This applies to the general public, as well as to political and economic decision-makers. In Europe, and particularly in France, the image of algae conveyed by the media is often negative. They are associated with ecological disturbances, such as green tides in Brittany or brown tides in the French overseas departments of Guadeloupe and Martinique. In the latter case, it is funny to note that sargassum (*Sargassum fluitans*, *Sargassum natans*) belongs to the same algal family as the sargassum (*Sargassum muticum*) used by traditional Chinese and Korean medicines.

This devalued image of algae in the Western collective unconscious dates back to the emergence of the algae tide phenomenon. The Romans, a Mediterranean people unfamiliar with tides and algae, discovered this resource during their conquest of north-western Europe. The algae that washed up with every tide or storm was regarded by them as waste washed up by the sea. This negative perception of algae by Roman society is probably the origin of the Western world's poor perception of this marine resource. This negative image was also reinforced during periods of famine suffered by Western populations.

In a book with the evocative title *Les nourritures des affamés : décrire et penser les aliments de substitution au haut Moyen Âge*[10], the author writes, "Some went so far as to eat algae" (Vandenberg 2015).

During food shortages, coastal populations in Norway and Iceland relied on algae. Icelandic populations in particular used algae belonging to the genera *Alaria* and *Rhodymenia* (*Palmaria*) (Sauvageau 1920). *Rhodymenia palmata*, or *Palmaria palmata*, is abundant on the North Atlantic coast, and has a protein content of up to 25 or even 35% of dry weight, depending on harvesting and cultivation conditions. This last point makes this species a valuable source of protein in times of shortage or famine.

For Western consumers, however, the image of algae is associated with periods of famine, and is likened by some to a kind of "Jerusalem artichoke" from the sea.

It is generally not highly valued, given the marginal nature of its consumption and its food history in Europe.

In Japan, algae consumption is also linked to periods of war and famine. Japanese soldiers ate dried rice, legume pastes, dried fish and dehydrated algae, such as kombu (*Laminaria japonica*), arame (*Eisenia bicyclis*), hijiki (*Hizikia fusiformis*) and wakame (*Undaria pinnatifida*). In times of famine, these same algae were considered sources of sustenance. Despite this context, the image of algae among Japanese consumers was never altered, as this marine resource was also consumed outside periods of war and famine. For this reason, the vision of algae as a foodstuff is totally opposed to that held by Western consumers.

10 This title can be translated as *Foods for the Hungry: Describing and Thinking About Substitute Foods in the Early Middle Ages*.

In addition to these aspects, which are linked to a devaluing perception of algae in the Western world, algae are relatively little studied compared to terrestrial plants. The number of publications on macroalgae, the main organisms used in traditional medicine, represents 8.45% of the total number of publications relating to the plant world (see Figure 4.11).

This difference is even more marked for the main models of agronomic or agri-food interest. Indeed, publications relating to soy (*Glycine max*) total 3,620,000 scientific articles, compared with 27,900 for kombu (*Laminaria japonica*), the main algae exploited by humans worldwide (see Figure 4.12).

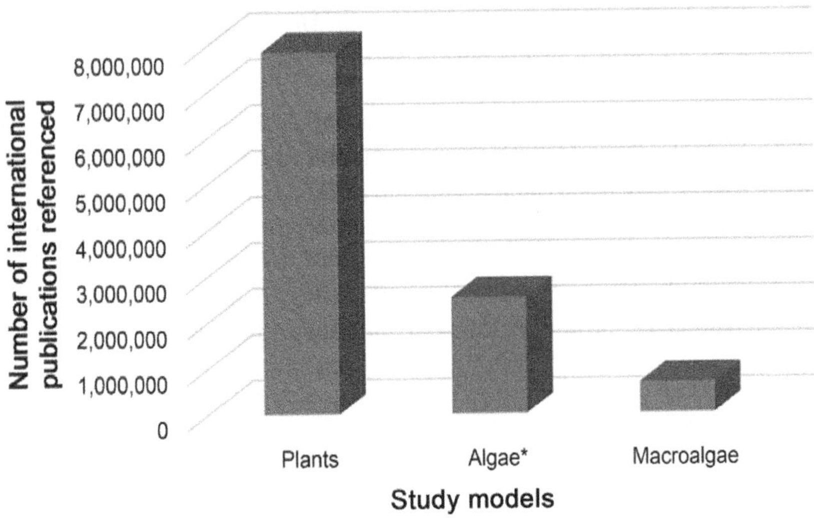

Figure 4.11. *Number of international scientific publications relating to different study models of plant origin (algae*: microalgae + macroalgae) (from Google Scholar (2024))*

The visibility of a scientific field can be indirectly assessed by a bibliometric criterion: the impact factor (IF) of scientific journals dealing with the field. The top 2% are journals with an IF of over 10, and the top 10% are scientific journals with an impact factor of over 4. At present, no scientific journal dealing with algology in the broadest sense appears in the top 2% (see Figure 4.13). On the other hand, several journals related to the biology of terrestrial plants feature in this ranking, including *Trends in Plant Sciences*, which boasts an IF of 20.5. Two other journals, *Molecular Plant* and *Plant Cell*, are also on this list, with IFs of 17.37 and 11.6 respectively.

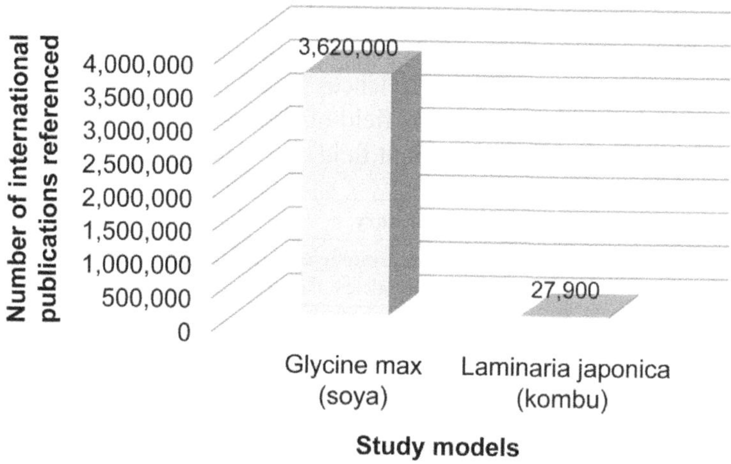

Figure 4.12. *Number of international scientific publications relating to soy* (Glycine max) *and kombu* (Laminaria japonica) *(from Google Scholar (2024))*

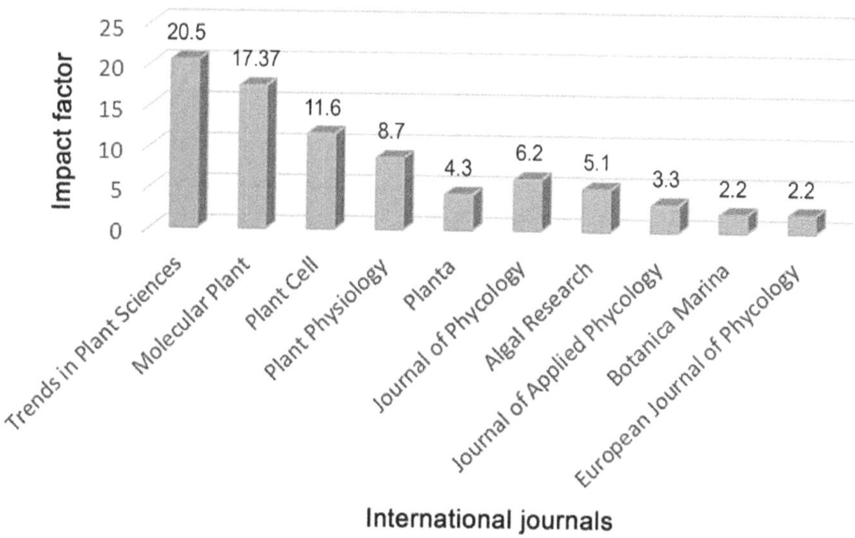

Figure 4.13. *Impact factors of various scientific journals dealing with plants or algae (according to Google, on 2022–2023 data)*

By contrast, at least two algae-related journals are in the top 10%, namely *Algal Research* (IF 5.1) and *Journal of Phycology* (IF 6.2).

The IF is the best-known and most widely used bibliometric criterion for assessing the reputation of scientific journals. It is not, however, an absolute criterion of journal quality or excellence. In the case of algology-related journals, it simply indicates that this field of research is less well known and less worked on than others in the plant field.

Teams	Laboratory	Affiliated organization
Génétique des algues	Laboratoire de biologie intégrative des modèles marins	CNRS Roscoff-Sciences Sorbonne Université
Morphogénèse des macroalgues	Laboratoire de biologie intégrative des modèles marins	CNRS Roscoff-Sciences Sorbonne Université
Glycobiologie marine	Laboratoire de biologie intégrative des modèles marins	CNRS Roscoff-Sciences Sorbonne Université
Biotechnologie : valorisation de molécules marines	Laboratoire de biotechnologie et chimie marines	Université de Bretagne Sud (Vannes)
Axe : découverte, étude et valorisation de molécules et d'ingrédients d'origine marine pour la santé et la biotechnologie	LEMAR	Université de Bretagne Occidentale CNRS IRD Ifremer (Brest)
Équipe valorisation pour une aquaculture durable (EVAD)	ISOMER	Nantes Université
Remote Sensing Benthic Ecology and Ecotoxicology (RSBE2)	ISOMER	Nantes Université
Écologie marine et biodiversité	Institut méditerranéen d'océanologie	Université d'Aix-Marseille

Table 4.3. *Main French academic structures involved in the research and development of algae (macroalgae)*

In France, the number of organizations and research teams working on algae (macroalgae) is relatively limited (see Table 4.3). The laboratories concerned are mainly located on the Atlantic coast, and more particularly in Brittany, a French region historically involved in the use of algal resources. Ifremer, a research organization dedicated to the sustainable exploitation of marine resources, has not carried out research into the biology and development of macroalgae for some 20 years, and therefore no longer has its own laboratory dedicated to the subject. However, in the late 1980s, this organization was at the forefront of the development of brown alga *Undaria pinnatifida* off the island of Ouessant. By way of indication, INRAE, the research organization in charge of agronomic research, boasts 273 research laboratories, service and experimentation units.

These findings on national research show that algae is not given much consideration in public policy. This is due to a professional sector that is less structured and less visible than the agricultural sector, as well as to lower societal expectations on the part of the public.

More generally, in the Western world, algae are the victims of an often biased and not always positive perception. This often applies to macroalgae, the subject of this book. On the other hand, microalgae used as nutritional supplements or cell factories in the biotechnological field enjoy a generally favorable image with the general public (Fleurence 2021).

This perception, which applies to macroalgae, is often the result of a past and present that are insufficiently informed, and therefore misunderstood.

its own laboratory dedicated to the subject. However, in the late [illegible], this organization was at the forefront of the development of brown algal [illegible]

[illegible]

Conclusion

The sea is often regarded as a haven of molecules of therapeutic interest, and therefore as a source of tomorrow's medicines. However, the development of a drug from a natural substance is governed by at least four rules, which can be stated as follows:

– The compound concerned must have a biological activity of therapeutic interest.

– Its mechanism of action must be fully understood and distinguished from that of similar products.

– The compound and the resource from which it is derived must be available.

– Its industrial application must be protected by intellectual property that is difficult to circumvent.

In the case of algal molecules, the first rule is always easy to respect. The second is not always, and depends on the individual case. On the other hand, the third principle is much more difficult to achieve, given the biochemical variations affecting algae, particularly macroalgae, which are the basis of traditional medicines. The last rule is also difficult to respect, especially when dealing with natural molecules. Only protection of the obtaining process or the application of the molecule can be envisaged in this field.

In much more general terms, the development of a therapeutic molecule is empirically based on the rule of 5 stated by Lipinski in 1997. This rule states the following principles:

– no more than five donor hydrogen bonds;

– no more than 10 hydrogen bond acceptors;

– a molecular weight of less than 500 daltons;

– an octanol–water partition coefficient not exceeding 5.

This rule, also known as Pfizer's rule of 5, is made up of values that are all multiples of 5.

Most of the algal molecules described above do not respect this rule, as is the case for many natural products.

For all the above reasons, the development of drugs based on algal molecules, outside current applications (dressings, biomaterials), is far from an easy road.

Against this backdrop, the use of algae in human health could be envisaged on a preventive rather than curative basis. This could be achieved by introducing algae into our diet, as is the case in Asia.

This final thought is not, in itself, innovative, if we refer to Hippocrates' famous phrase: "Let food be thy medicine and medicine be thy food".

References

Abdul, Q.A., Choi, R.J., Jung, H.A., Choi, J.S. (2016). Health benefit of fucosterol from marine algae: A review. *Journal of the Science of Food and Agriculture*, 96(6), 1856–1866.

Admassu, H., Gasmalla, M.A.A., Yang, R., Zhao, W. (2018). Bioactive peptides derived from seaweed protein and their health benefits: Antihypertensive, antioxidant, and antidiabetic properties. *Journal of Food Science*, 83(1), 6–16.

Ahan, R.E., Hanifehnezhad, A., Kehribar, E.S., Oguzoglu, T.C., Foldes, K., Özçelik, C.E., Filazi, N., Öztop, S., Palaz, F., Önder, S. et al. (2022). A highly potent SARS-CoV-2 blocking lectin protein. *ACS Infectious Diseases*, 8(7), 1253–1264.

Airanthi, M.W.A., Hosokawa, M., Miyashita, K. (2011). Comparative antioxidant activity of edible Japanese brown seaweeds. *Journal of Food Science*, 76(1), C104–C111.

Allen, N.E., Sauvaget, C., Roddam, A.W., Appleby, P., Nagano, J., Suzuki, G., Key, T., Koyama, K. (2004). A prospective study of diet and prostate cancer in Japanese men. *Cancer Causes & Control*, 15, 911–920.

Almutairi, F.M., Adams, G.G., Kök, M.S., Lawson, C.J., Gahler, R., Wood, S., Foster, T.J., Rowe, A.J., Harding, S.E. (2013). An analytical ultracentrifugation based study on the conformation of lambda carrageenan in aqueous solution. *Carbohydrate Polymers*, 97(1), 203–209.

Álvarez-Viñas, M., Souto, S., Flórez-Fernández, N., Torres, M.D., Bandín, I., Domínguez, H. (2021). Antiviral activity of carrageenans and processing implications. *Marine Drugs*, 19(8), 437.

André, R., Pacheco, R., Bourbon, M., Serralheiro, M.L. (2021). Brown algae potential as a functional food against hypercholesterolemia. *Foods*, 10(2), 234.

Anggadiredja, J.T. (2009). Ethnobotany study of seaweed diversity and its utilization in Warambadi, Panguhalodo areas of East Sumba District. *Jurnal Teknologi Lingkungan*, 10(3), 297–310.

Arasaki, S. and Arasaki, T. (1985). *Basse calorie, haute nutrition. Les légumes de mer. Comment être et paraître en forme*. Guy Trédaniel, Paris.

Arijón, M., Ponce, N.M., Solana, V., Dellatorre, F.G., Latour, E.A., Stortz, C.A. (2021). Monthly fluctuations in the content and monosaccharide composition of fucoidan from *Undaria pinnatifida* sporophylls from northern Patagonia. *Journal of Applied Phycology*, 33(4), 2433–2441.

Arumugam, G., Rajendran, R., Khaleelullah, N.S., Ramanathan, S. (2019). Anti-candidal and anti-virulence efficiency of selected seaweeds against azole resistance *Candida albicans*. *Biocatalysis and Agricultural Biotechnology*, 20, 101195.

Balansa, W. (2023). Exploring metabolites of green algae *Caulerpa* spp. to discover putative inhibitors of Dpp-4. A new antidiabetic target protein: Molecular docking compounds from marine green algae. *Jurnal Ilmiah PLATAX*, 11(2), 333–346.

Bei, H., Guang-Ce, W., Chen-Kui, Z., Zhen-gang, L. (2002). The experimental research of R-phycoerythrin subunits on cancer treatment: A new photosensitizer in PDT. *Cancer Biotherapy and Radiopharmaceuticals*, 17(1), 35–42.

Bellik, Y., Boukraâ, L., Alzahrani, H.A., Bakhotmah, B.A., Abdellah, F., Hammoudi, S.M., Iguer-Ouada, M. (2012). Molecular mechanism underlying anti-inflammatory and anti-allergic activities of phytochemicals: An update. *Molecules*, 18(1), 322–353.

Besednova, N.N., Andryukov, B.G., Zaporozhets, T.S., Kuznetsova, T.A., Kryzhanovsky, S.P., Ermakova, S.P., Galkina, I.V., Shchelkanov, M.Y. (2022). Molecular targets of brown algae phlorotannins for the therapy of inflammatory processes of various origins. *Marine Drugs*, 20(4), 243.

Brown, E.M., Allsopp, P.J., Magee, P.J., Gill, C.I., Nitecki, S., Strain, C.R., McSorley, E.M. (2014). Seaweed and human health. *Nutrition Reviews*, 72(3), 205–216.

Buck, C.B., Thompson, C.D., Roberts, J.N., Müller, M., Lowy, D.R., Schiller, J.T. (2006). Carrageenan is a potent inhibitor of papillomavirus infection. *PLoS Pathogens*, 2(7), e69.

Bustillos, J.V. (1856). *Elementos de farmacia aplicada a la medicina obra aprobada i adoptada para la enseñanza por la Universidad de Chile compuesta*. Imprenta del Frerrocarri, Santiago.

Catarino, M.D., Silva, A., Cruz, M.T., Mateus, N., Silva, A.M., Cardoso, S.M. (2020). Phlorotannins from *Fucus vesiculosus*: Modulation of inflammatory response by blocking NF-κB signaling pathway. *International Journal of Molecular Sciences*, 21(18), 6897.

Chandini, S.K., Ganesan, P., Bhaskar, N. (2008). In vitro antioxidant activities of three selected brown seaweeds of India. *Food Chemistry*, 107(2), 707–713.

Chatterji, A., Kassim, Z., Hassan, A., Therwath, A., Shaharom, F. (2010). Marine living resources in the practice of traditional medicine. *Journal Coastal Environment*, 1(1), 41–52.

Cho, J.Y., Kang, J.Y., Khan, M.N.A., Park, N.H., Kim, S.K., Hong, Y.K. (2007). Anti-inflammatory activities of *Undaria pinnatifida* and *Laminaria japonica* (Phaeophyta). *Fisheries and Aquatic Sciences*, 10(3), 127–132.

Colliec, S., Fischer, A.M., Tapon-Bretaudiere, J., Boisson, C., Durand, P., Jozefonvicz, J. (1991). Anticoagulant properties of a fucoidan fraction. *Thrombosis Research*, 64(2), 143–154.

Conte, É. and Payri, C. (2002). La consommation des algues en Polynésie française : premiers résultats d'une enquête. *Journal de la Société des Océanistes*, 114/115, 165–172.

Cotas, J., Leandro, A., Monteiro, P., Pacheco, D., Figueirinha, A., Gonçalves, A.M., da Silva, G.J., Pereira, L. (2020). Seaweed phenolics: From extraction to applications. *Marine Drugs*, 18(8), 384.

Cotas, J., Pacheco, D., Gonçalves, A.M., Silva, P., Carvalho, L.G., Pereira, L. (2021). Seaweeds' nutraceutical and biomedical potential in cancer therapy: A concise review. *Journal of Cancer Metastasis and Treatment*, 7, 13.

Craigie, J.S. (2011). Seaweed extract stimuli in plant science and agriculture. *Journal of Applied Phycology*, 23, 371–393.

Czelusta, K.L., Zafar, W., Khan, F., Banu, S., Aslam, H., Shah, A. (2018). Vitamins, minerals, and herbs: Their effects on neurocognitive disorders. *Directions in Psychiatry*, 38(2), 149–164.

Darcy-Vrillon, B. (1993). Nutritional aspects of the developing use of marine macroalgae for the human food industry. *International Journal of Food Sciences and Nutrition*, 44, 523–525.

De Martin, S., Gabbia, D., Carrara, M., Ferri, N. (2018). The brown algae *Fucus vesiculosus* and *Ascophyllum nodosum* reduce metabolic syndrome risk factors: A clinical study. *Natural Product Communications*, 13(12). doi: 10.1177/1934578X1801301228.

De Sousa, A.P.A., Torres, M.R., Pessoa, C., de Moraes, M.O., Rocha Filho, F.D., Alves, A.P.N.N., Costa-Lotufo, L.V. (2007). In vivo growth-inhibition of Sarcoma 180 tumor by alginates from brown seaweed *Sargassum vulgare*. *Carbohydrate Polymers*, 69(1), 7–13.

Denis, C., Morançais, M., Li, M., Deniaud, E., Gaudin, P., Wielgosz-Collin, G., Barnathan, G., Jaouen, P., Fleurence, J. (2010). Study of the chemical composition of edible red macroalgae *Grateloupia turuturu* from Brittany (France). *Food Chemistry*, 119(3), 913–917.

Derosa, G., Cicero, A. F., D'Angelo, A., Maffioli, P. (2019). *Ascophyllum nodosum* and *Fucus vesiculosus* on glycemic status and on endothelial damage markers in dysglicemic patients. *Phytotherapy Research*, 33(3), 791–797.

Dillehay, T.D., Ramírez, C., Pino, M., Collins, M.B., Rossen, J., Pino-Navarro, J.D. (2008). Monte Verde: Seaweed, food, medicine, and the peopling of South America. *Science*, 320(5877), 784–786.

Dumilag, R.V. and Javier, R.F. (2022). Ethnobotany of medicinal seaweeds of Ilocos Norte, Philippines. *Philippine Journal of Science*, 151, 1135–1156.

Echave, J., Otero, P., Garcia-Oliveira, P., Munekata, P.E., Pateiro, M., Lorenzo, J.M., Simal-Gandara, J., Prieto, M.A. (2022). Seaweed-derived proteins and peptides: Promising marine bioactives. *Antioxidants*, 11(1), 176.

Endo, M., Hori, M., Mihara, T., Ozaki, H., Oikawa, T., Odaguchi, H., Hanawa, T. (2017). *Zingiberis Siccatum Rhizoma*, the active component of the Kampo formula Daikenchuto, induces anti-inflammatory actions through α7 nicotinic acetylcholine receptor activation. *Neurogastroenterology & Motility*, 29(12), e13139.

Fischer, A.M., Tapon-Bretaudiere, J., Sternberg, C., Durand, P., Jozefonvicz, J. (1991). Biological activities of fucans extracted from. *Drugs of the Future*, 16(6), 539–545.

Fleurence, J. (1999). Seaweed proteins: Biochemical, nutritional aspects and potential uses. *Trends in Food Science & Technology*, 10(1), 25–28.

Fleurence, J. (2018). *Les algues alimentaires : bilan et perspectives*. Lavoisier Tec & Doc, Paris.

Fleurence, J. (2021). *Microalgae: From Future Food to Cellular Factory*. John Wiley & Sons, Hoboken.

Fleurence, J. (2022). Biostimulant potential of seaweed extracts derived from *Laminaria* and *Ascophyllum nodosum*. In *Biostimulants: Exploring Sources and Applications*, Ramawat, N. and Bhardwaj, V. (eds). Springer, Singapore, 31–49.

Fleurence, J. (2024). The seaweeds: A traditional and an alternative source of proteins for the human and animal nutrition. In *Recent Advances in Seaweed Biotechnology*, Trivedi, N., Reddy, C.R.K., Critchley, A.T. (eds). Springer, Berlin.

Fleurence, J. and Gall, E.A. (2016). Antiallergic properties. In *Seaweed in Health and Disease Prevention*, Fleurence, J. and Levine, I. (eds). Academic Press, Cambridge, 389–406.

Fleurence, J., Gutbier, G., Mabeau, S., Leray, C. (1994). Fatty acids from 11 marine macroalgae of the French Brittany coast. *Journal of Applied Phycology*, 6, 527–532.

Frediansyah, A. (2021). The antiviral activity of iota-, kappa-, and lambda-carrageenan against COVID-19: A critical review. *Clinical Epidemiology and Global Health*, 12, 100826.

Freile-Pelegrin, Y. and Morales, J.L. (2004). Antibacterial activity in marine algae from the coast of Yucatan, Mexico. *Botanica Marina*, 47(2), 140–146.

Galland-Irmouli, A.V., Pons, L., Lucon, M., Villaume, C., Mrabet, N.T., Guéant, J.L., Fleurence, J. (2000). One-step purification of R-phycoerythrin from the red macroalga *Palmaria palmata* using preparative polyacrylamide gel electrophoresis. *Journal of Chromatography B: Biomedical Sciences and Applications*, 739(1), 117–123.

Geng, L., Zhang, Q., Suo, Q., Wang, J., Wang, Y., Wang, C., Wu, N. (2023). Inhibitory activity of a sulfated oligo-porphyran from *Pyropia yezoensis* against SARS-CoV-2. *Carbohydrate Polymers*, 299, 120173.

Gerasimenko, N.I., Busarova, N.G., Martyyas, E.A. (2012). Composition of lipids from *Fucus evanescens* (Seas of Okhotsk and Japan) and biological activity of lipids and photosynthetic pigments. *Chemistry of Natural Compounds*, 48, 742–747.

Gilboa-Garber, N., Citronbaum, R., Levene, C., Sela, R. (1988). H blood group detection by the L-fucose binding lectin of the green marine alga *Ulva lactuca*. *Developmental & Comparative Immunology*, 12(4), 695–705.

Girard, J.P., Marion, C., Liutkus, M., Boucard, M., Rechencq, E., Vidal, J.P., Rossi, J.C. (1988). Hypotensive constituents of marine algae; 1. Pharmacological studies of laminine. *Planta Medica*, 54(03), 193–196.

Gómez-Zorita, S., González-Arceo, M., Trepiana, J., Eseberri, I., Fernández-Quintela, A., Milton-Laskibar, I., Aguirre, L., Gonzalez, M., Portillo, M.P. (2020). Anti-obesity effects of macroalgae. *Nutrients*, 12(8), 2378.

Gonzalez, M.E., Alarcón, B., Carrasco, L. (1987). Polysaccharides as antiviral agents: Antiviral activity of carrageenan. *Antimicrobial Agents and Chemotherapy*, 31(9), 1388–1393.

Gupta, S. and Abu-Ghannam, N. (2011). Bioactive potential and possible health effects of edible brown seaweeds. *Trends in Food Science & Technology*, 22(6), 315–326.

Han, Y., Zhang, L., Yu, X., Wang, S., Xu, C., Yin, H., Wang, S. (2019). Retracted Article: Alginate oligosaccharide attenuates α2, 6-sialylation modification to inhibit prostate cancer cell growth via the Hippo/YAP pathway. *Cell Death & Disease*, 10(5), 374.

Harb, T.B. and Chow, F. (2022). Anti-HIV activity of methanolic and aqueous extracts of fifteen materials of beach-cast macroalgae: Valorization of underused waste biomass. *Applied Phycology*, 3(1), 236–246.

Hayashi, K., Walde, P., Miyazaki, T., Sakayama, K., Nakamura, A., Kameda, K., Masuda, S., Umakoshi, H., Kato, K. (2012). Active targeting to osteosarcoma cells and apoptotic cell death induction by the novel lectin *Eucheuma serra* agglutinin isolated from a marine red alga. *Journal of Drug Delivery*, Article ID 842785. doi: 10.1155/2012/842785.

Hellio, C., Bremer, G., Pons, A.M., Le Gal, Y., Bourgougnon, N. (2000). Inhibition of the development of microorganisms (bacteria and fungi) by extracts of marine algae from Brittany, France. *Applied Microbiology and Biotechnology*, 54, 543–549.

Hermans, M., Akoègninou, A., van der Maesen, J. (2004). Medicinal plants used to treat malaria in southern Benin. *Economic Botany*, 58(1), S239–S252.

Hoang, M.H., Jia, Y., Jun, H.J., Lee, J.H., Lee, B.Y., Lee, S.J. (2012). Fucosterol is a selective liver X receptor modulator that regulates the expression of key genes in cholesterol homeostasis in macrophages, hepatocytes, and intestinal cells. *Journal of Agricultural and Food Chemistry*, 60(46), 11567–11575.

Holdt, S.L. and Kraan, S. (2011). Bioactive compounds in seaweed: Functional food applications and legislation. *Journal of Applied Phycology*, 23, 543–597.

Hu, X., Jiang, X., Hwang, H., Liu, S., Guan, H. (2004). Antitumour activities of alginate-derived oligosaccharides and their sulphated substitution derivatives. *European Journal of Phycology*, 39(1), 67–71.

Hu, X., Jiang, X., Aubree, E., Boulenguer, P., Critchley, A.T. (2006). Preparation and in vivo. Antitumor activity of κ-carrageenan oligosaccharides. *Pharmaceutical Biology*, 44(9), 646–650.

Hudson, J.B., Kim, J.H., Lee, M.K., DeWreede, R.E., Hong, Y.K. (1998). Antiviral compounds in extracts of Korean seaweeds: Evidence for multiple activities. *Journal of Applied Phycology*, 10, 427–434.

Huh, G.W., Lee, D.Y., In, S.J., Lee, D.G., Park, S.Y., Yi, T.H., Baek, N.I. (2012). Fucosterols from *Hizikia fusiformis* and their proliferation activities on osteosarcoma-derived cell MG63. *Journal of the Korean Society for Applied Biological Chemistry*, 55, 551–555.

Hung, L.D., Hirayama, M., Ly, B.M., Hori, K. (2015). Purification, primary structure, and biological activity of the high-mannose N-glycan-specific lectin from cultivated *Eucheuma denticulatum*. *Journal of Applied Phycology*, 27, 1657–1669.

Iacoviello, L., Zito, F., Rago, L., Di Castelnuovo, A., De Curtis, A., Zappacosta, B., de Gaetano, G., Donati, M.B., Cerletti, C. (2013). Prolonged administration of *Ascophyllum nodosum* to healthy human volunteers and cardiovascular risk. *Nutrafoods*, 12, 137–144.

Iso, H. and Kubota, Y. (2007). Nutrition and disease in the Japan collaborative cohort study for evaluation of cancer (JACC). *Asian Pacific Journal of Cancer Prevention*, 8(Suppl), 35–80.

Ito, K. and Tsuchiya, Y. (1977). Differential fatty acid composition of some marine algae associated with their habitat depths. *Tohoku Journal of Agricultural Research*, 28(3), 145–150.

James, C.A., Welham, S., Rose, P. (2023). Edible algae allergenicity – a short report. *Journal of Applied Phycology*, 35(1), 339–352.

Ji, Y.B., Ji, C.F., Yue, L. (2014). Study on human promyelocytic leukemia HL-60 cells apoptosis induced by fucosterol. *Bio-Medical Materials and Engineering*, 24(1), 845–851.

Jiménez-Escrig, A., Jiménez-Jiménez, I., Pulido, R., Saura-Calixto, F. (2001). Antioxidant activity of fresh and processed edible seaweeds. *Journal of the Science of Food and Agriculture*, 81(5), 530–534.

Jin, W., He, X., Wu, W., Bao, Y., Wang, S., Cai, M., Zhang, W., Wang, C., Zhang, F., Linhardt, R.J. et al. (2020). Structural analysis of a glucoglucuronan derived from laminarin and the mechanisms of its anti-lung cancer activity. *International Journal of Biological Macromolecules*, 163, 776–787.

Joel, C.H., Sutopo, C.C., Prajitno, A., Su, J.H., Hsu, J.L. (2018). Screening of angiotensin-I converting enzyme inhibitory peptides derived from *Caulerpa lentillifera*. *Molecules*, 23(11), 3005.

Jung, H.A., Islam, M.N., Lee, C.M., Oh, S.H., Lee, S., Jung, J.H., Choi, J.S. (2013a). Kinetics and molecular docking studies of an anti-diabetic complication inhibitor fucosterol from edible brown algae *Eisenia bicyclis* and *Ecklonia stolonifera*. *Chemico-Biological Interactions*, 206(1), 55–62.

Jung, H.A., Jin, S.E., Ahn, B.R., Lee, C.M., Choi, J.S. (2013b). Anti-inflammatory activity of edible brown alga *Eisenia bicyclis* and its constituents fucosterol and phlorotannins in LPS-stimulated RAW264. 7 macrophages. *Food and Chemical Toxicology*, 59, 199–206.

Kadam, S.U., Tiwari, B.K., O'Donnell, C.P. (2015). Extraction, structure and biofunctional activities of laminarin from brown algae. *International Journal of Food Science & Technology*, 50(1), 24–31.

Kang, J.Y., Khan, M.N.A., Park, N.H., Cho, J.Y., Lee, M.C., Fujii, H., Hong, Y.K. (2008). Antipyretic, analgesic, and anti-inflammatory activities of the seaweed *Sargassum fulvellum* and *Sargassum thunbergii* in mice. *Journal of Ethnopharmacology*, 116(1), 187–190.

Karuppusamy, S., Rajauria, G., Fitzpatrick, S., Lyons, H., McMahon, H., Curtin, J., Tiwari, B., O'Donnell, C. (2022). Biological properties and health-promoting functions of laminarin: A comprehensive review of preclinical and clinical studies. *Marine Drugs*, 20(12), 772.

Keleszade, E., Patterson, M., Trangmar, S., Guinan, K.J., Costabile, A. (2021). Clinical efficacy of brown seaweeds *Ascophyllum nodosum* and *Fucus vesiculosus* in the prevention or delay progression of the metabolic syndrome: A review of clinical trials. *Molecules*, 26(3), 714.

Kendel, M., Couzinet-Mossion, A., Viau, M., Fleurence, J., Barnathan, G., Wielgosz-Collin, G. (2013). Seasonal composition of lipids, fatty acids, and sterols in the edible red alga *Grateloupia turuturu*. *Journal of Applied Phycology*, 25, 425–432.

Kendel, M., Wielgosz-Collin, G., Bertrand, S., Roussakis, C., Bourgougnon, N., Bedoux, G. (2015). Lipid composition, fatty acids and sterols in the seaweeds *Ulva armoricana*, and *Solieria chordalis* from Brittany (France): An analysis from nutritional, chemotaxonomic, and antiproliferative activity perspectives. *Marine Drugs*, 13(9), 5606–5628.

Khan, M.N., Choi, J.S., Lee, M.C., Kim, E., Nam, T.J., Fujii, H., Hong, Y.K. (2008). Anti-inflammatory activities of methanol extracts from various seaweed species. *Journal of Environmental Biology*, 29(4), 465–469.

Khotimchenko, Y.S. (2010). The antitumor properties of nonstarch polysaccharides: Carrageenans, alginates, and pectins. *Russian Journal of Marine Biology*, 36, 401–412.

Khotimchenko, M., Tiasto, V., Kalitnik, A., Begun, M., Khotimchenko, R., Leonteva, E., Bryukhovetskiy, I., Khotimchenko, Y. (2020). Antitumor potential of carrageenans from marine red algae. *Carbohydrate Polymers*, 246, 116568.

Kim, Y.M., Kim, D.S., Choi, Y.S. (2004). Anticoagulant activities of brown seaweed extracts in Korea. *Korean Journal of Food Science and Technology*, 36(6), 1008–1013.

Kim, K.H., Kim, Y.W., Kim, H.B., Lee, B.J., Lee, D.S. (2006). Anti-apoptotic activity of laminarin polysaccharides and their enzymatically hydrolyzed oligosaccharides from *Laminaria japonica*. *Biotechnology Letters*, 28, 439–446.

Kloareg, B. (1991). Structure et propriétés d'échange des parois cellulaires des Algues brunes. Implications écophysiologiques. *Bulletin de la Société Botanique de France. Actualités Botaniques*, 138(3/4), 305–318.

Kobayashi, K. (2016). A review on palliative care of stomach cancer with Kampo medication in Japan. Acupuncture and Chinese Medicine PhD Thesis, University of Anaheim, Anaheim.

Kumar, S. and Barros, U. (2020). Purification and partial characterization of a haemagglutinin from *Ulva fasciata*. *Current Science*, 118(4), 621.

Kwon, M.J. and Nam, T.J. (2006). Porphyran induces apoptosis related signal pathway in AGS gastric cancer cell lines. *Life Sciences*, 79(20), 1956–1962.

Le Tutour, B., Benslimane, F., Gouleau, M.P., Gouygou, J.P., Saadan, B., Quemeneur, F. (1998). Antioxidant and pro-oxidant activities of the brown algae, *Laminaria digitata*, *Himanthalia elongata*, *Fucus vesiculosus*, *Fucus serratus* and *Ascophyllum nodosum*. *Journal of Applied Phycology*, 10, 121–129.

Leong, Y.K. and Chang, J.S. (2024). Proteins and bioactive peptides from algae: Insights into antioxidant, anti-hypertensive, anti-diabetic and anti-cancer activities. *Trends in Food Science & Technology*, 145, 104352.

Li, C., Tang, T., Du, Y., Jiang, L., Yao, Z., Ning, L., Zhu, B. (2023). Ulvan and Ulva oligosaccharides: A systematic review of structure, preparation, biological activities and applications. *Bioresources and Bioprocessing*, 10(1), 66.

Liu, L., Heinrich, M., Myers, S., Dworjanyn, S.A. (2012). Towards a better understanding of medicinal uses of the brown seaweed *Sargassum* in Traditional Chinese Medicine: A phytochemical and pharmacological review. *Journal of Ethnopharmacology*, 142(3), 591–619.

Liu, Y., Geng, L., Zhang, J., Wang, J., Zhang, Q., Duan, D., Zhang, Q. (2018). Oligo-porphyran ameliorates neurobehavioral deficits in parkinsonian mice by regulating the PI3K/Akt/Bcl-2 pathway. *Marine Drugs*, 16(3), 82.

Lomartire, S. and Gonçalves, A.M. (2022). Antiviral activity and mechanisms of seaweeds bioactive compounds on enveloped viruses – A review. *Marine Drugs*, 20(6), 385.

Lopes, D., Rey, F., Leal, M.C., Lillebø, A.I., Calado, R., Domingues, M.R. (2021). Bioactivities of lipid extracts and complex lipids from seaweeds: Current knowledge and future prospects. *Marine Drugs*, 19(12), 686.

Maeno, Y., Terada, R., Kotaki, Y., Cho, Y., Konoki, K., Yotsu-Yamashita, M. (2019). Possible biosynthetic products and metabolites of kainic acid from the red alga *Digenea simplex* and their biological activity. *Journal of Natural Products*, 82(6), 1627–1633.

Maliki, I.M., Misson, M., Teoh, P.L., Rodrigues, K.F., Yong, W.T.L. (2022). Production of lectins from marine algae: Current status, challenges, and opportunities for non-destructive extraction. *Marine Drugs*, 20(2), 102.

Mao, Z., Shen, X., Dong, P., Liu, G., Pan, S., Sun, X., Hu, H., Pan, L., Huang, J. (2019). Fucosterol exerts antiproliferative effects on human lung cancer cells by inducing apoptosis, cell cycle arrest and targeting of Raf/MEK/ERK signalling pathway. *Phytomedicine*, 61, 152809.

Matsukawa, R., Dubinsky, Z., Kishimoto, E., Masaki, K., Masuda, Y., Takeuchi, T., Chihara, M., Yamamoto, E., Niki, E., Karube, I. (1997). A comparison of screening methods for antioxidant activity in seaweeds. *Journal of Applied Phycology*, 9, 29–35.

McDermid, K.J., Martin, K.J., Haws, M.C. (2019). Seaweed resources of the Hawaiian Islands. *Botanica Marina*, 62(5), 443–462.

McLaughlin, C.M., Sharkey, S.J., Harnedy-Rothwell, P., Parthsarathy, V., Allsopp, P.J., McSorley, E.M., Fitzgerald, R.J., O'Harte, F.P. (2020). Twice daily oral administration of *Palmaria palmata* protein hydrolysate reduces food intake in streptozotocin induced diabetic mice, improving glycaemic control and lipid profiles. *Journal of Functional Foods*, 73, 104101.

Meinita, M.D.N., Harwanto, D., Tirtawijaya, G., Negara, B.F.S.P., Sohn, J.H., Kim, J.S., Choi, J.S. (2021). Fucosterol of marine macroalgae: Bioactivity, safety and toxicity on organism. *Marine Drugs*, 19(10), 545.

Michikawa, T., Inoue, M., Shimazu, T., Sawada, N., Iwasaki, M., Sasazuki, S., Ymaji, T., Tsugane, S., Japan Public Health Center-based Prospective Study Group (2012). Seaweed consumption and the risk of thyroid cancer in women: The Japan Public Health Center-based Prospective Study. *European Journal of Cancer Prevention*, 21(3), 254–260.

Miclon, M., Courtot, É., Guégnard, F., Lenhof, O., Boudesocque-Delaye, L., Matard-Mann, M., Nyvall Collén, P., Castagnone Sereno, P., Neveu, C. (2023). The brown alga presents an anthelmintic activity on all developmental stages of the parasitic nematode *Heligmosomoides polygyrus bakeri*. *Pathogens*, 12(4), 540.

Mookerjea, S. and Hawkins, W.W. (1958). The antilipaemic activity of laminarin sulphate. *Canadian Journal of Biochemistry and Physiology*, 36(3), 261–268.

Mori, T., O'Keefe, B.R., Sowder, R.C., Bringans, S., Gardella, R., Berg, S., Cochran, P., Turpin, J.A., Buckheit, R.W. Jr., McMahon, J.B. et al. (2005). Isolation and characterization of griffithsin, a novel HIV-inactivating protein, from the red alga *Griffithsia* sp. *Journal of Biological Chemistry*, 280(10), 9345–9353.

Motoo, Y., Seki, T., Tsutani, K. (2011). Traditional Japanese medicine, Kampo: Its history and current status. *Chinese Journal of Integrative Medicine*, 17(2), 85–87.

Nakazawa, S. (1974). Antitumor effect of water-extracts from marine algae. *Chemotherapy*, 22, 1435–1442.

Nang, H.Q. and Dinh, N.H. (1998). The seaweed resources of Vietnam. In *Seaweed Resources of the World*, Critchley, T. and Ohno, M. (eds). Japan International Cooperation Agency, Tokyo, 62–78.

Necas, J. and Bartosikova, L. (2013). Carrageenan: A review. *Veterinarni Medicina*, 58(4), 187–205.

Nestler, G. (2002). Traditional Chinese medicine. *Medical Clinics*, 86(1), 63–73.

Neyrinck, A.M., Mouson, A., Delzenne, N.M. (2007). Dietary supplementation with laminarin, a fermentable marine β (1–3) glucan, protects against hepatotoxicity induced by LPS in rat by modulating immune response in the hepatic tissue. *International Immunopharmacology*, 7(12), 1497–1506.

Noda, H., Amano, H., Arashima, K., Nisizawa, K. (1990). Antitumor activity of marine algae. *Hydrobiologia*, 204, 577–584.

Noguchi, T., Matsui, T., Miyazawa, K., Asakawa, M., Iijima, N., Shida, Y., Fuse, M., Hosaka, Y., Kirigaya, C., Watabe, K. et al. (1994). Poisoning by the red alga "ogonori" (*Gracilaria verrucosa*) on the Nojima Coast, Yokohama, Kanagawa Prefecture, Japan. *Toxicon*, 32(12), 1533–1538.

Ochiai, Y., Katsugari, T., Hashimoto, K. (1987). Proteins in three seaweeds, "aosa" *Ulva lactuca*, "arame" *Eisenia bicyclis*, and "makusa" *Gelidium amansii*. *Nippon Suisan Gakkaishi*, 53(6), 1051–1055.

O'Connor, K. (2017). *Seaweed: A Global History*. Reaktion Books, London.

O'Keefe, B.R., Giomarelli, B., Barnard, D.L., Shenoy, S.R., Chan, P.K.S., McMahon, J.B., Palmer, K.E., Barnett, B.W., Meyerholz, D.K., Wohlford-Lenane, C.L. et al. (2010). Broad-spectrum in vitro activity and in vivo efficacy of the antiviral protein griffithsin against emerging viruses of the family Coronaviridae. *Journal of Virology*, 84(5), 2511–2521.

Pádua, D., Rocha, E., Gargiulo, D., Ramos, A.A. (2015). Bioactive compounds from brown seaweeds: Phloroglucinol, fucoxanthin and fucoidan as promising therapeutic agents against breast cancer. *Phytochemistry Letters*, 14, 91–98.

Pan, Q., Chen, M., Li, J., Wu, Y., Zhen, C., Liang, B. (2013). Antitumor function and mechanism of phycoerythrin from *Porphyra haitanensis*. *Biological Research*, 46(1), 87–95.

Paradis, M.E., Couture, P., Lamarche, B. (2011). A randomised crossover placebo-controlled trial investigating the effect of brown seaweed (*Ascophyllum nodosum* and *Fucus vesiculosus*) on postchallenge plasma glucose and insulin levels in men and women. *Applied Physiology, Nutrition, and Metabolism*, 36(6), 913–919.

Park, J.H., Kim, Y.H., Kwon, T.K., Park, J.H., Woo, J.S. (2016). Antiviral composition comprising a carrageenan as an active ingredient. Korea Patent KR20160003964A, 12.

Pereira, L. and Valado, A. (2021). The seaweed diet in prevention and treatment of the neurodegenerative diseases. *Marine Drugs*, 19(3), 128.

Pérez, R. (1997). *Ces algues qui nous entourent : conception actuelle, rôle dans la biosphere, utilisations, culture*. Éditions Quæ, Paris.

Pérez-Lloréns, J.L., Critchley, A.T., Cornish, M.L., Mouritsen, O.G. (2023). Saved by seaweeds (II): Traditional knowledge, home remedies, medicine, surgery, and pharmacopoeia. *Journal of Applied Phycology*, 35, 2049–2068.

Perumal, K.M., Gandhi, D., Rajasekaran, M.B.S., Kudiyarasu, S., Renuka, R.R., Julius, A., Samrot, A.V., Narayanan, A.L. (2023). Inhibition of angiogenesis using laminarin a natural polysaccharide from brown seaweeds – A review. *Biocatalysis and Agricultural Biotechnology*, 102947.

Qian, L., Zhou, Y., Ma, J.X. (2014). Hypolipidemic effect of the polysaccharides from *Porphyra yezoensis*. *International Journal of Biological Macromolecules*, 68, 48–49.

Qiu, Y., Jiang, H., Fu, L., Ci, F., Mao, X. (2021). Porphyran and oligo-porphyran originating from red algae Porphyra: Preparation, biological activities, and potential applications. *Food Chemistry*, 349, 129209.

Rani, V., Shakila, R.J., Jawahar, P., Srinivasan, A. (2017). Influence of species, geographic location, seasonal variation and extraction method on the fucoidan yield of the brown seaweeds of Gulf of Mannar, India. *Indian Journal of Pharmaceutical Sciences*, 79, 65–71.

Sanjeewa, K.K., Lee, W., Jeon, Y.J. (2018). Nutrients and bioactive potentials of edible green and red seaweed in Korea. *Fisheries and Aquatic Sciences*, 21(1), 1–11.

Sauvageau, C. (1920). *Utilisation des algues marines*, volume 1. Librairie Octave Doin, Paris.

Shi, Z., Takeuchi, T., Nakanishi, Y., Kato, T., Beck, K., Nagata, R., Kageyama, T., Ito, A., Ohno, H., Satoh-Takayama, N. (2022). A Japanese herbal formula, Daikenchuto, alleviates experimental colitis by reshaping microbial profiles and enhancing group 3 innate lymphoid cells. *Frontiers in Immunology*, 13. doi: 10.3389/fimmu.2022.903459.

Shimazu, T., Kuriyama, S., Hozawa, A., Ohmori, K., Sato, Y., Nakaya, N., Nishino, Y., Tsubono, Y., Tsuji, I. (2007). Dietary patterns and cardiovascular disease mortality in Japan: A prospective cohort study. *International Journal of Epidemiology*, 36(3), 600–609.

Skriptsova, A.V. (2016). Seasonal variations in the fucoidan content of brown algae from Peter the Great Bay, Sea of Japan. *Russian Journal of Marine Biology*, 42, 351–356.

Springer, G.F., Wurzel, H.A., McNeal, G.M. Jr., Ansell, N.J., Doughty, M.F. (1957). Isolation of anticoagulant fractions from crude fucoidin. *Proceedings of the Society for Experimental Biology and Medicine*, 94(2), 404–409.

Srikong, W., Bovornreungroj, N., Mittraparparthorn, P., Bovornreungroj, P. (2017). Antibacterial and antioxidant activities of differential solvent extractions from the green seaweed *Ulva intestinalis*. *ScienceAsia*, 43(2), 88–95.

Stoyneva-Gärtner, M.P. and Uzunov, B.A. (2017). Seaweeds used as remedies in Bulgaria. *Oceanography & Fisheries Open Access Journal*, 1(4), 75–76.

Sugiura, Y., Katsuzaki, H., Imai, K., Amano, H. (2021). The anti-allergic and anti-inflammatory effects of phlorotannins from the edible brown algae, *Ecklonia* sp. and *Eisenia* sp. *Natural Product Communications*, 16(12), 1934578X211060924.

Sun, S., Xu, X., Sun, X., Zhang, X., Chen, X., Xu, N. (2019). Preparation and identification of ACE inhibitory peptides from the marine macroalga *Ulva intestinalis*. *Marine Drugs*, 17(3), 179.

Szekalska, M., Puciłowska, A., Szymańska, E., Ciosek, P., Winnicka, K. (2016). Alginate: Current use and future perspectives in pharmaceutical and biomedical applications. *International Journal of Polymer Science*, Article ID 7697031. doi: 10.1155/2016/7697031.

Tang, H.F., Yi, Y.H., Yao, X.S., Xu, Q.Z., Zhang, S.Y., Lin, H.W. (2002). Bioactive steroids from the brown alga *Sargassum carpophyllum*. *Journal of Asian Natural Products Research*, 4(2), 95–101.

Teas, J., Vena, S., Cone, D.L., Irhimeh, M. (2013). The consumption of seaweed as a protective factor in the etiology of breast cancer: Proof of principle. *Journal of Applied Phycology*, 25, 771–779.

Terada, R. and Watanebe, Y. (2016). Seaweeds and seagrasses in the Amami Islands: Biodiversity and utilization. In *The Amani Islands: Culture, Society, Industry and Nature*, Kawai, K., Terada, R., Kuwahara, S. (eds). Hokuto Shobo Publishing, Tokyo, 107–115.

Thangam, R., Senthilkumar, D., Suresh, V., Sathuvan, M., Sivasubramanian, S., Pazhanichamy, K., Gorlagunta, P.K., Kannan, S., Gunasekaran, P., Rengasamy, R., Sivaraman, J. (2014). Induction of ROS-dependent mitochondria-mediated intrinsic apoptosis in MDA-MB-231 cells by glycoprotein from Codium decorticatum. *Journal of Agricultural and Food Chemistry*, 62(15), 3410–3421.

Thomson, A.W. and Fowler, E. (1981). Carrageenan: A review of its effects on the immune system. *Agents and Actions*, 11, 265–273.

Ulagesan, S., Nam, T.J., Choi, Y.H. (2021). Extraction and purification of R-phycoerythrin alpha subunit from the marine red algae *Pyropia yezoensis* and its biological activities. *Molecules*, 26(21), 6479.

Unschuld, P.U. (2018). *Traditional Chinese Medicine: Heritage and Adaptation*. Columbia University Press, Columbia.

Van Wyk, B.E. (2008). A review of Khoi-San and Cape Dutch medical ethnobotany. *Journal of Ethnopharmacology*, 119(3), 331–341.

Vandenberg, V. (2015). Les nourritures des affamés : décrire et penser les aliments de substitution au haut Moyen Âge. *L'alimentazione nell'alto Medioevo. Pratiche, simboli, ideologie*, 1, 447–466.

Vera, J., Castro, J., Gonzalez, A., Moenne, A. (2011). Seaweed polysaccharides and derived oligosaccharides stimulate defense responses and protection against pathogens in plants. *Marine Drugs*, 9(12), 2514–2525.

Viebke, C., Borgström, J., Piculell, L. (1995). Characterisation of kappa-and iota-carrageenan coils and helices by MALLS/GPC. *Carbohydrate Polymers*, 27(2), 145–154.

Vo, T.S., Ngo, D.H., Kim, S.K. (2012). Marine algae as a potential pharmaceutical source for anti-allergic therapeutics. *Process Biochemistry*, 47(3), 386–394.

Vonthron-Sénécheau, C. (2016). Medicinal properties: Antibiotic, tonic, and antiparasitic properties. In *Seaweed in Health and Disease Prevention*, Fleurence, J. and Levine, I. (eds). Academic Press, Cambridge, 369–388.

Vonthron-Sénécheau, C., Kaiser, M., Devambez, I., Vastel, A., Mussio, I., Rusig, A.M. (2011). Antiprotozoal activities of organic extracts from French marine seaweeds. *Marine Drugs*, 9(6), 922–933.

Wang, H., Ooi, E.V., Ang, P.O. Jr. (2008). Antiviral activities of extracts from Hong Kong seaweeds. *Journal of Zhejiang University Science B*, 9(12), 969–976.

Wang, W., Zhang, P., Hao, C., Zhang, X.E., Cui, Z. Q., Guan, H.S. (2011). In vitro inhibitory effect of carrageenan oligosaccharide on influenza A H1N1 virus. *Antiviral Research*, 92(2), 237–246.

Wang, W., Wang, S.X., Guan, H.S. (2012). The antiviral activities and mechanisms of marine polysaccharides: An overview. *Marine Drugs*, 10(12), 2795–2816.

Weiner, M.L. (1991). Toxicological properties of carrageenan. *Agents and Actions*, 32(1/2), 46–51.

Widyaningsih, W., Salamah, N., Maulida, F.Q. (2016). The effects of ethanolic extract of green algae (*Ulva lactuca* L.) on blood cholesterol levels in male rats induced by a high fat diet. *JKKI: Jurnal Kedokteran dan Kesehatan Indonesia*, 7, 181–186.

Wijesinghe, W.A.J.P. and Jeon, Y.J. (2012). Biological activities and potential industrial applications of fucose rich sulfated polysaccharides and fucoidans isolated from brown seaweeds: A review. *Carbohydrate Polymers*, 88(1), 13–20.

Wong, T.Y., Preston, L.A., Schiller, N.L. (2000). Alginate lyase: Review of major sources and enzyme characteristics, structure-function analysis, biological roles, and applications. *Annual Reviews in Microbiology*, 54(1), 289–340.

Xing, M., Cao, Q., Wang, Y., Xiao, H., Zhao, J., Zhang, Q., Aiguo, J., Song, S. (2020). Advances in research on the bioactivity of alginate oligosaccharides. *Marine Drugs*, 18(3), 144.

Yamada, T., Ogamo, A., Saito, T., Watanabe, J., Uchiyama, H., Nakagawa, Y. (1997). Preparation and anti-HIV activity of low-molecular-weight carrageenans and their sulfated derivatives. *Carbohydrate Polymers*, 32(1), 51–55.

Yamada, T., Ogamo, A., Saito, T., Uchiyama, H., Nakagawa, Y. (2000). Preparation of O-acylated low-molecular-weight carrageenans with potent anti-HIV activity and low anticoagulant effect. *Carbohydrate Polymers*, 41(2), 115–120.

Yamamoto, I., Takahashi, M., Tamura, E., Maruyama, H. (1982). Antitumor activity of crude extracts from edible marine algae against L-1210 leukemia. *Botanica Marina*, 25(9), 455–457.

Yamori, Y., Miura, A., Taira, K. (2001). Implications from and for food cultures for cardiovascular diseases: Japanese food, particularly Okinawan diets. *Asia Pacific Journal of Clinical Nutrition*, 10(2), 144–145.

Yang, Y.J., Nam, S.J., Kong, G., Kim, M.K. (2010). A case–control study on seaweed consumption and the risk of breast cancer. *British Journal of Nutrition*, 103(9), 1345–1353.

Yao, Z., Wu, H., Zhang, S., Du, Y. (2014). Enzymatic preparation of κ-carrageenan oligosaccharides and their anti-angiogenic activity. *Carbohydrate Polymers*, 101, 359–367.

Yasuzawa, T., Mima, A., Ueshima, S. (2019). Antithrombotic effect of oral administration of mozuku (*Cladosiphon okamuranus*, brown seaweed) extract in rat. *Journal of Nutritional Science and Vitaminology*, 65(2), 171–176.

Zargarzadeh, M., Amaral, A.J., Custódio, C.A., Mano, J.F. (2020). Biomedical applications of laminarin. *Carbohydrate Polymers*, 232, 115774.

Zemke-White, W. Lindsey and Ohno, M. (1999). World seaweed utilization: An end-of-century summary. *Journal of Applied Phycology*, 11, 369–376.

Index

A, C

D, E

F, G

H, K, L

M, O, P

R, S

T, U

Other titles from

ISTE

in

Agriculture, Food Science and Nutrition

2024

CHERIET Foued, MAUREL Carole, AMADIEU Paul, HANNIN Hervé
Wine Management and Marketing 2: Responses of the Industry to Crises and New Expectation

FAVENNEC Jean-Pierre
Geopolitics and Energy Transition 2: From One Continent to Another, Contrasting Situations

KOUBAA Mayssa, HAYKAL AMMAR Mohamed, DHOUIB Diala, MNEJJA Sirine
Optimization in the Agri-Food Supply Chain: Recent Studies

2023

FLEURENCE Joël
Algae in Agrobiology: Realities and Perspectives

2020

CHERIET Foued, MAUREL Carole, AMADIEU Paul, HANNIN Hervé
Wine Management and Marketing: Opportunities for Companies and Challenges for the Industry

2019

YANNOU-LE BRIS Gwenola, SERHAN Hiam, DUCHAÎNE Sibylle, FERRANDI Jean-Marc, TRYSTRAM Gilles
Ecodesign and Ecoinnovation in the Food Industries

2016

JEANTET Romain, CROGUENNEC Thomas, SCHUCK Pierre, BRULÉ Gérard
Handbook of Food Science and Technology 1: Food Alteration and Food Quality
Handbook of Food Science and Technology 2: Food Process Engineering and Packaging
Handbook of Food Science and Technology 3: Food Biochemistry and Technology